AF688306

IRINA RODICA RABEJA

DIODES AND TRANSISTORS

essentials

IRINA RODICA RABEJA

DIODES AND TRANSISTORS

essentials

National Library of Australia Cataloguing-in-Publication entry

Creator: Rabeja, Irina Rodica Olga, author.

Title: Diodes and transistors / Irina Rodica Rabeja.

ISBN: 9780977509874 (paperback)

Subjects: Diodes.
Transistors.
Electronics.

Publisher Irina Rabeja
Sydney Australia
2026

CONTENT

PART I

DIODES

THE DIODE

The diode was the first electronic device that revolutionized the science and the engineering in the year 1904, when was realized for the first time in the laboratory by the English savant John Ambroise Fleming as a vacuum glass tube with two electrodes inside performing the detection function.

About the year 1883, the renowned American inventor Thomas Alva Edison tried to improve the electric bulb composed by socket, filament and the protecting glass bulb. Introducing in the glass bulb a small metal plate and connecting between that and the filament a milliamperemeter, he observed the presence of an electric current when the filament was under voltage and incandescent.
This phenomenon was named after its discoverer the "Edison Effect".

John Ambroise Fleming was the first who realised the Edison Effect's importance.
He found out that the electrons emitted by the incandescent cathode spread in the space between cathode and anode forming a spatial electrical charge.
In the new circuit the small metal plate was named "anode" the filament was named "cathode" and the glass tube with vacuum containing two electrodes, the anode and the cathode, was named "diode".
Applying between the anode and the cathode a voltage with the positive polarity on anode, the electrons move from the cathode to the anode, so an electric current appears in the diode, see FIGURE 1; changing the polarity of the voltage, the electric current disappears because between the anode and the cathode appears an electric field which stops the electrons to move. So the diode permits the passing of an electric current when the voltage between the anode and the cathode is positive and stops it when that voltage is negative (the cathode is considered reference electrode).

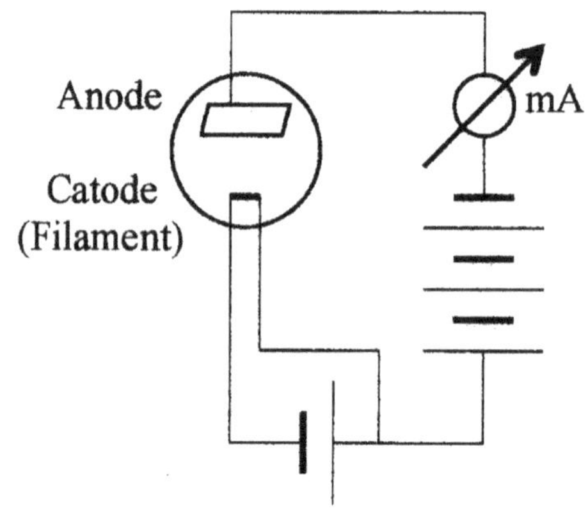

FIGURE 1 Edison effect development

Symbolically the diode is a commutator or switch ON/OFF; the state is determined by the polarity of the voltage.

A semiconductor is a solid whose electrical conductivity is in between that of a metal and that of an insulator. After the year 1948 the electronic devices are manufactured with germanium or silicon semiconductor crystals whose properties are determined by the forces between the electrons and the atomic nuclei.

Germanium (Ge) is a grey-white metalloid element of the group IV of the Periodic Table of Chemical Elements. It is by-product of the zinc ores or the combustion of certain coals. Silicon (Si) is a grey non-metallic element of the group IV of the Periodic Table.
It is found only in combinations such as silicate and silica, being the second most abundant element in the Earth's crust (27.7% of Earth's crust's weight).

The pure germanium crystal is an insulator from the electrical viewpoint, a dielectric. But if it is contaminated (process known as *doping*) with phosphor atoms, arsenic atoms or stibium/antimony atoms, which are penta-valent elements, it becomes a conductor. These impurities are called *donors* because they donate electrons to the semiconductor and the electric conductibility that appears is of type **n** (by moving **n**egative electricity carriers = electrons).

The electric conductibility can be obtained too by increasing artificially the number of "holes" created by electrons which leave. In this case in the crystal are introduced other impurities as boron atoms, aluminium atoms, gallium atoms or indium atoms, which are tri-valent elements, that create "holes" capturing (accepting) electrons.
These impurities are called *acceptors* and the electric conductibility that appears is of type **p** (by moving **p**ositive electricity carriers = "holes").

A crystal of germanium or silicon with two parts, type **n** and type **p**, side-by-side is called **pn** junction (the joint between a **p** region and a **n** region formes the **pn** junction).

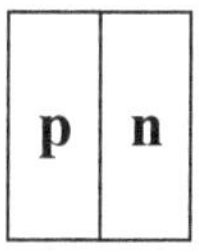

The **pn** junction was discovered by the American engineer Russell Ohl in the year 1940. Across the joint surface appears a zone, the *depletion* zone, with *spatial electric charge* missing the mobile electric charge carriers, consequence of the diffusion of electrons from the **n** region to the **p** region and the diffusion of "holes" from the **p** region to the **n** region. The spatial electric charge creates an electric field which hinder further diffusion until is struck the equilibrium.
A semiconductor crystal diode is in essence a **pn** junction.
The graphical display depicting complex non-linear relationships of the electronic circuits in *static/dynamic* (direct/variable) regimes is called the *characteristic curve*.
The symbol of the semiconductor diode and its ideal static characteristic curve $I = f(V)$ or I-V are drawn in FIGURE 2.
The diode permits the running of an electric current I if the positive electrode of a voltage source V (for ex. a battery) is connected to the **p** region and the negative electrode of the same source is connected to the **n** region (the diode is like a short circuit). The diode does not permit the running of an electric current I if the positive electrode of a voltage source V (for ex. a battery) is connected to the **n** region and the negative electrode of the same source is connected to the **p** region (the diode is like an open circuit).

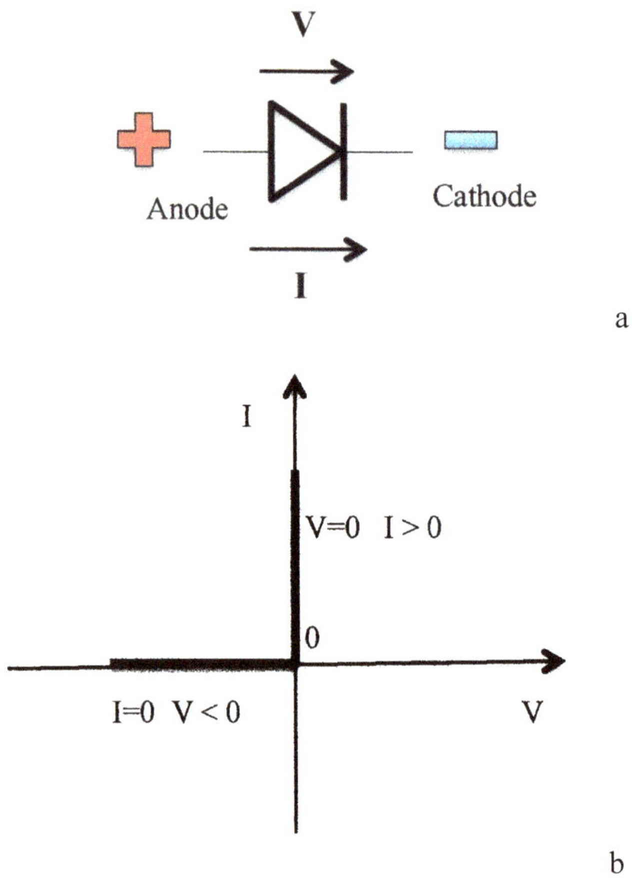

FIGURE 2
a. Diode symbol
b. I-V static characteristic of the ideal diode

The two above described states of a diode are called *forward conduction* respectively *reverse conduction* and their electric current *forward current* and *reverse current.* The analysis of the static electric behaviour of the **pn** diode led to the formula:

$$I = I_R\{\exp\,[eV/(kT)] - 1\}$$

I through-diode current (A)

V on-diode voltage (V)

I_R I in saturation reverse conduction or *residual current* ($|V| >> kT/e$)

e electron electrical charge $1.602177 \cdot 10^{-19}$ C

k Boltzman constant $1.380658 \cdot 10^{-23}$ J/K

T temperature in Kelvin degrees K

$V_T = kT/e$ thermal voltage

$V_T = 0.025$ V for Ge at 290 K or 17 °C

$e/kT = 0.04$ S/mA $= 40\ V^{-1}$ for Ge at 290 K or 17 °C

Forward conduction $I = I_R \exp[eV/(kT)] = I_R \exp(V/V_T)$

$\exp[eV/(kT)] = \exp(V/V_T) >> 1 \quad (V > 0 \quad V >> V_T)$

Reverse conduction $I = -I_R$

$\exp[eV/(kT)] = \exp(V/V_T) << 1 \quad (V < 0 \quad |V| >> V_T)$

The diode static characteristic curve $I = f(V)$ is drawn in FIGURE 3.

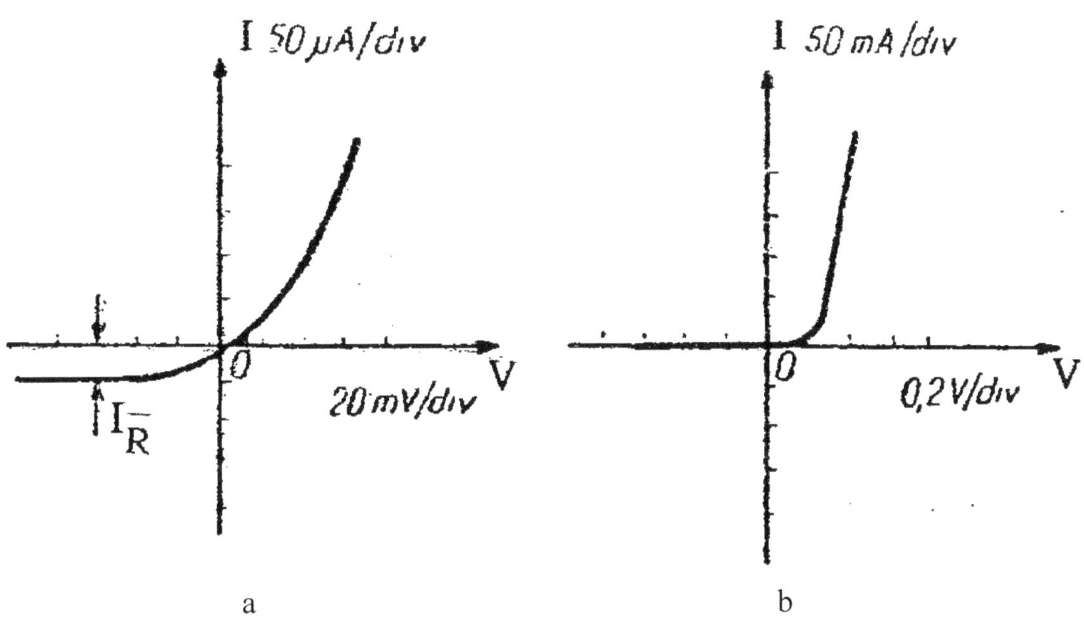

FIGURE 3 I-V static characteristic of a pn diode
a. around V= 0 b. complete characteristic

The experimental static characteristic curve of a diode $I = f(V)$ given in FIGURE 4 is different from the characteristic given by the previous formulas.

In *reverse conduction* there are two regions: (I) and (II)

(I) current increases slowly by increasing reverse voltage

(II) current has a sudden increase at a reverse voltage called breakdown voltage

In *forward conduction* appear four regions: (III) (IV) (V) (VI)

(III) current has exponential variation with voltage $I \approx \exp[eV/(zkT)]$ z=1...2

(IV) current has parabolical variation with voltage $I \approx V^2$

(V) current has exponential variation with voltage $I \approx \exp[eV/(ckT)]$ c=const

(VI) current varies linear with voltage and is limited by load electrical resistance

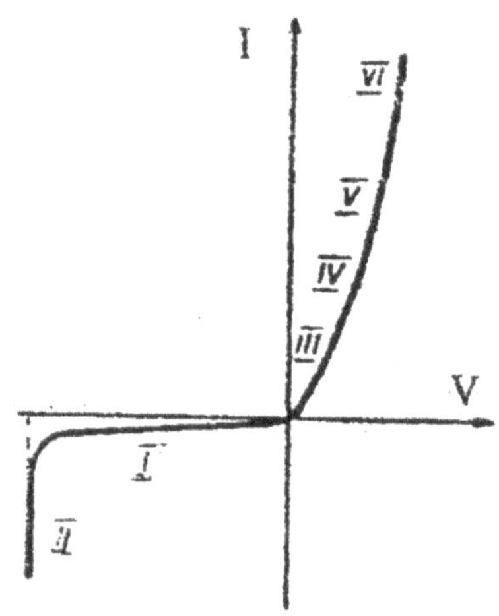

FIGURE 4 I-V experimental static characteristic of pn junction

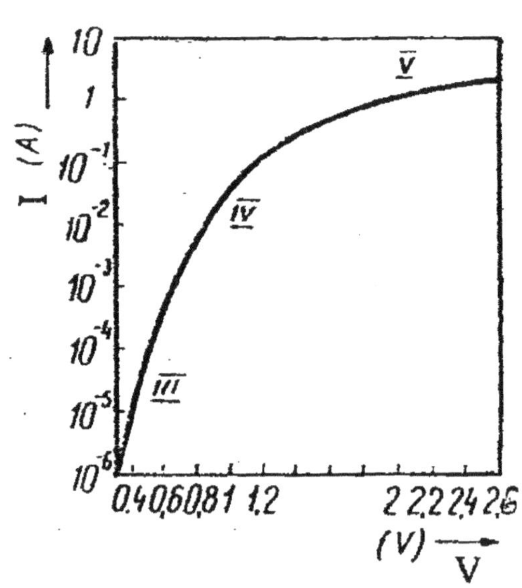

FIGURE 5. I-V experimental static characteristic of pn junction in forward conduction in semi-logarithmic scale

In FIGURE 5 is presented the experimental static characteristic curve of a **pn** junction in forward conduction using a semi-logarithmic scale.

It is known that overheating conductors and semiconductors, those can emit electrons. For each temperature there is generated a certain proportion of electron & "hole" *pairs. The increasing number of free electrons and "holes" when rising the temperature increases the electric conductivity of the semiconductors, quite different from the conductors where the electric conductivity decreases.*

In general if the temperature increases the diode electric current increases.
FIGURE 6 presents the I-V static characteristic curve for **pn** junction for two different temperatures T2 >T1.

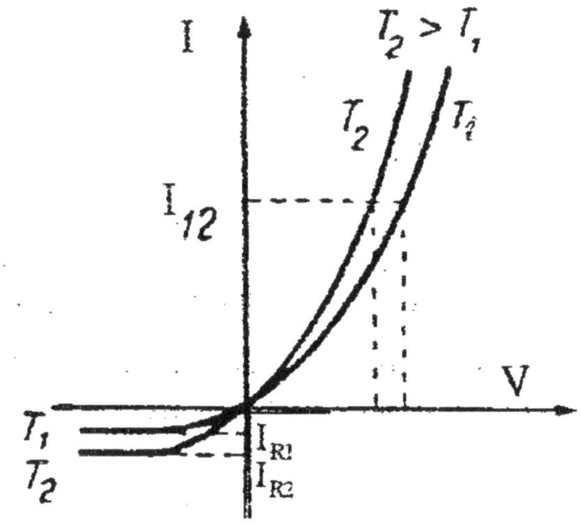

FIGURE 6. I-V static characteristic of pn junction for two different temperatures $T_2 > T_1$

If the forward current is constant the voltage on **pn** junction decreases if the temperature increases.
The residual current I_R doubles at every 7.5 ˚C for the **pn** junctions manufactured with germanium and for those manufactured with silicon doubles at every 6.5 ˚C.
But I_R of the silicon diodes is much smaller than that of the germanium diodes, so the temperature influence is lower for the silicon diodes.

There is a diode with a guaranteed low residual current named the *picoampere diode*.

The diode can be used in a great variety of electronic circuits, doing various tasks (depending on the application) making use of its many features. Having in mind this idea the diode was manufactured in a diversity of types that in the last half of century have been improving and multiplying at a fascinating speed.
For practical purposes further are presented the rectifier diode, the Zener diode, the commutation diode, the detection diode, the varicap diode, the tunnel diode, the IMPATT diode, the double-base diode, the controlled diode, the photodiode, the light-emitting diode.
For the diode, in general the catalogues present:

I_F diode electric current in forward conduction, over its maximum value the diode is destroyed (A)
I_R diode electric current in saturation reverse conduction (residual current)
I_{FRM} maximum repetitive I_F
I_{FSM} maximum surge (peak) I_F
V_F voltage on diode in forward conduction (V)
V_R voltage on diode in reverse conduction, over its maximum value the diode is destroyed
V_{RRM} maximum repetitive V_R
V_{RSM} maximum surge (peak) V_R

P_D electric power dissipation, over its maximum value the diode is destroyed (W)
T_j diode junction temperature, over its maximum value the diode is destroyed (˚C)
$T_{jstorage}$ range of allowable temperatures for storing a diode (un-powered)
T_A ambient temperature (referring to the cooling environment temperature)
R_{th} thermal resistance (˚C/W)
r junction electrical resistance (Ω)
C_j junction capacitance intrinsic to **pn** junction due to depletion zone (pF)
t_{rr} reverse recovery time for diode to "turn off" (s) ideally 0 s/usually μs…ns
f operating frequency (Hz)

The theory of the **pn** junction diode was developed in the year 1949 by the American physicist and inventor William Shockley and was instrumental for the invention of the bipolar junction transistor.

Diodes and transistors are individual electronic components or devices connected by electrical conductive wires with other individual electronic components or devices like resistors, capacitors, inductors and together they compose the electronic circuits.

The electronic components are divided in active and passive.

Components incapable of controlling electric current by means of another electrical signal are called passive components. Resistors, capacitors, inductors, transformers, and even diodes are all considered passive devices.

Components capable of controlling electric current by means of another electrical signal are called active components. Active components are capable of amplifying and generating signals. Transistors and also diodes presenting a negative dynamic resistance are considered active components.

In order for a circuit to be properly called electronic, it must contain at least one active component.

Since the rapid development of the modern electronics it is recommended to consult the latest components catalogues.

RECTIFIER DIODE

The rectifier diode is the diode used in the rectifying process.
The rectifying process is the conversion of the alternating electric current (AC) into the direct electric current (DC) by allowing the current to flow only in one direction using valves. The device used in this process is called the rectifier.
The electronic rectifier uses the rectifying feature of the vacuum diode or the semiconductor diode (**pn** junction).
The rectifier diode works in a narrow range of electrical waves frequencies generally around 50 Hz.
Special rectifiers use diodes able to convert down to 10-20 Hz alternating voltages.
The semiconductor rectifier diode is defined by the following parameters:

frequency range
maximum current in forward conduction
voltage on the diode in forward conduction
maximum voltage in reverse conduction
maximum power dissipation
maximum junction temperature
junction-capsule(case) thermal resistance

Some rectifiers have to work at voltages and currents that are beyond the working limits of the diodes. The solution is to use more diodes in series connection for high voltages as in FIGURE 7 or in parallel connection for high currents as in FIGURE 8.
In the series connection, for further protection, in parallel with the diodes there are resistors to equal the voltages on the diodes in reverse conduction and capacitors to equal the voltages on the diodes at the apparition of the reverse voltage or of the short super-voltages since if a diode is off faster it would support the entire reverse voltage.
In the parallel connection, for further protection, series with the diodes are resistors to equal the currents in forward conduction.

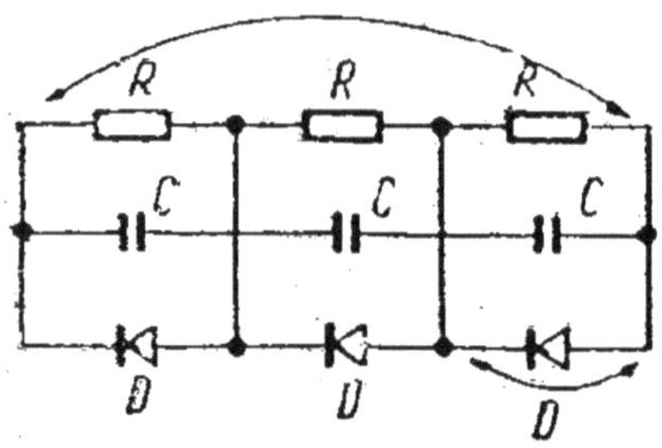

FIGURE 7
Series connection of the diodes

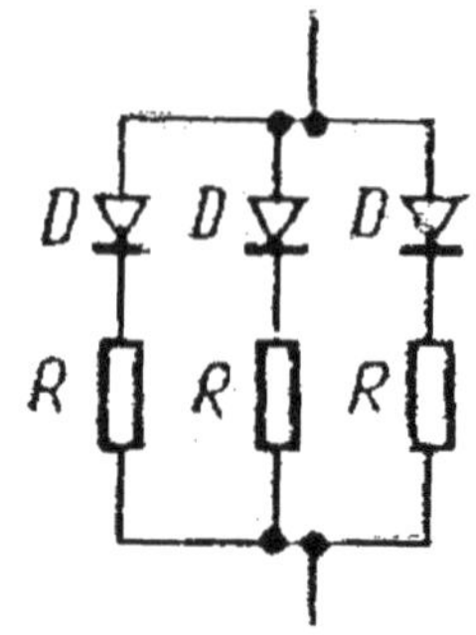

FIGURE 8
Parallel connection of the diodes

In FIGURE 9 there is presented a bridge rectifier with series diodes with:

(1) RC additional protection groups as R_1C_1 group at super-voltages appeared when the current is interrupted and R_2C_2 group to reduce super-voltages when the rectifier is connected at the main and the load is unconnected

(2) RC protection groups at cyclic super-voltages

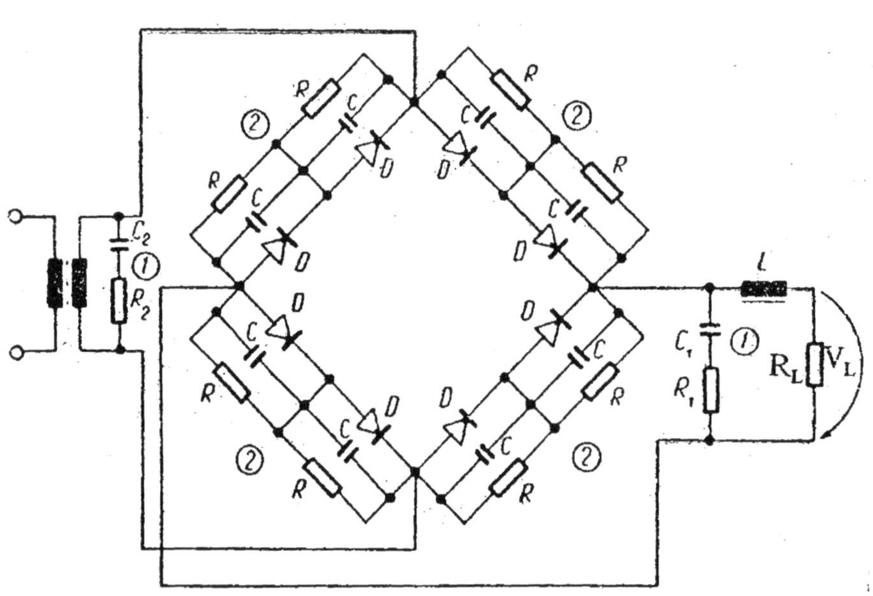

FIGURE 9 Monophase bridge rectifier with protection circuits

The power rectifier diodes (silicon) have nominal currents bigger than 20 A (even thousands A) and permitted reverse voltages up to 2000V, the voltage across a diode being 1.2 V.

By connecting the diodes in series to resist to higher voltages and in parallel to permit high rectified current the power of the rectifiers using silicon diodes can reach MW values (direct current sources for electrolysis, rectifiers for electric traction of heavy loads etc).

In the power rectifiers, in the time of the rectifying process on the rectifying diodes there is power dissipation shown in FIGURE 10 for voltages over 100 V.

That is in general 8 W/kW for germanium diodes (diagram 1) and 4-5 W/kW for silicon diodes (diagram 2).

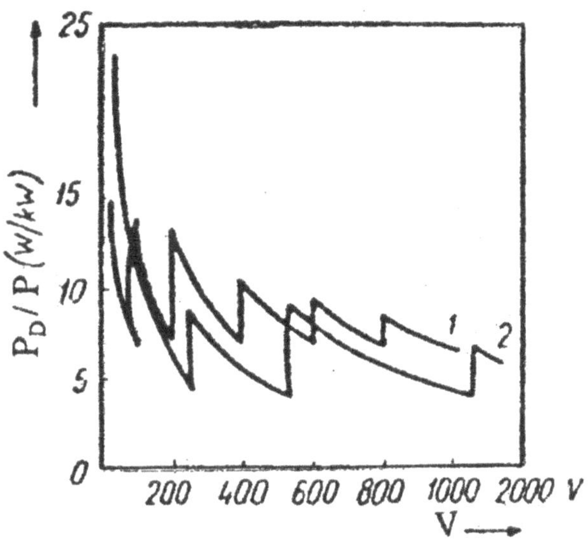

FIGURE 10
Power dissipation on the Ge(1) Si(2) diodes

The power dissipation on the diodes can heat them up and the overheat can destroy them. For thermal protection the diodes with high power dissipation, the power diodes, are fixed on heat sinks (metal radiators) that help to transmit the heat from the diode to the environment. The power diode is built to be fixed on a heat sink (metal radiator).

A vertical section through a power diode is presented in FIGURE 11.

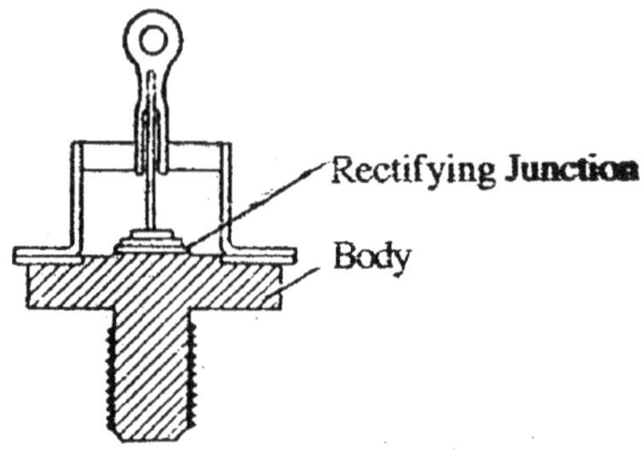

FIGURE 11
Vertical section through a power diode

The maximum power dissipated on the diode that is permitted is given by the formula:

$$P_D = (T_{jmax} - T_A) / R_{thJA} \qquad (W)$$

T_{jmax} maximum value of junction temperature, over it the junction is destroyed (˚C)

T_A ambient temperature (referring to the cooling environment temperature) (˚C)

R_{thJA} thermal resistance between **pn** junction and ambient (˚C/W)

The thermal resistance between **pn** junction and environment is the ratio of the temperature difference between the junction and the environment to the dissipated power on the diode at thermal equilibrium:

$$R_{thJA} = (T_j - T_A)/P_D \qquad (˚C/W)$$

The thermal resistance can be also expressed as a sum of other thermal resistances:

$$R_{thJA} = R_{thJC} + R_{thCR} + R_{thRA} + R_{th\,A} = R_{thJC} + R_{thCA}$$

$R_{thJC} = (T_j - T_c)/P_D$ junction-capsule thermal resistance

$R_{thCR} = (T_c - T_r)/P_D$ capsule-radiator thermal resistance

$R_{thRA} = (T_r - T_A)/P_D$ radiator-ambient thermal resistance

$R_{th\,A} = (T_{A1} - T_{A2})/P_D$ ambient thermal resistance

T_j junction temperature

T_c case(capsule) temperature

T_r radiator(heat sink) temperature

The junction-capsule thermal resistance R_{thJC} is the ratio of temperature difference between junction and capsule(case) to dissipated power at thermal equilibrium (given in the catalogues).

The capsule-radiator thermal resistance R_{thCR} is the ratio of the temperature difference between capsule(case) and heat sink (radiator) to dissipated power at thermal equilibrium; depends on the radiator finishing surfaces and pressure on the capsule; it is considered $\approx 1^{\circ}C/W$.

The radiator-ambient thermal resistance R_{thRA} is the ratio of the temperature difference between heat sink (radiator) and environment (close to radiator) to dissipated power at thermal equilibrium.

The ambient thermal resistance $R_{th\,A}$ is the ratio of the temperature difference between that close to radiator and average environment to dissipated power at thermal equilibrium; depends on the heat sink size and the movement of air and the big rectifiers use forced ventilation.

R_{thJC} depends on the diode manufacturing and can not be modified.

R_{thCA} can be decreased using heat sinks(radiators) with big surfaces.

The relationship between R_{thCA} and heat sink surface S is:

$$R_{thCA} = 1/(a_r S) \qquad (^{\circ}C/mW)$$

$a_r = 1.5\ mW/(cm^2\ ^{\circ}C)$

The heat sink surface can also be obtained from the diagrams of FIGURE 12 by knowing R_{thRA} ; the a,b,c diagrams are for a blackened Cu heat sink, as the d,e,f diagrams are for Cu heat sink with forced ventilation.

Depending on their amplitude and duration the electrical pulses could overload the diodes, pulses like that shown in FIGURE 13 where:

I_D average electric current through the working diode (A)

I_{DMax} electric current for which was designed the heat sink (A)

I_{dimp} super-current through diode (A)

t_K pulse duration (s)

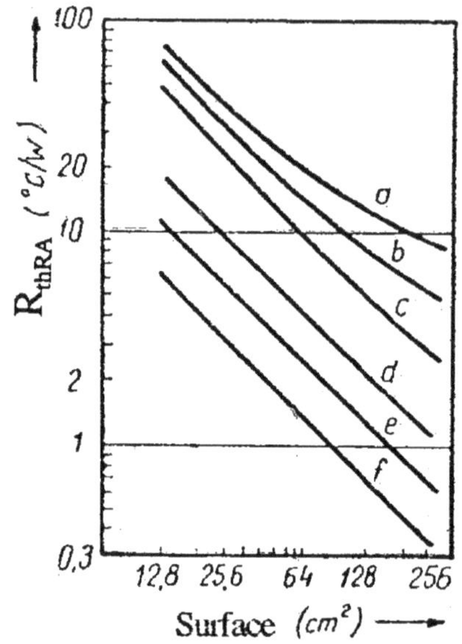

FIGURE 12

The influence of forced cooling on thermal resistance of heat sinks

- thermal resistance for blackened Cu heat sink with thicknesses (a)0.24mm (b)0.76mm (c)2.08mm for T_A=30°C T_r=100°C
- thermal resistance for Cu heat sink 2.08mm thick with forced cooling at (d)152m/min (e)305m/min (f)610m/min for T_A=30°C T_r=100°C

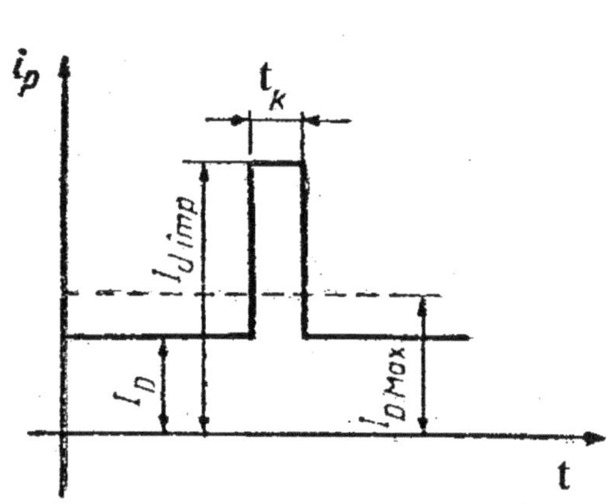

FIGURE 13 Overloading current pulse of diode

The permitted overload of the diodes depending on the heat sink and the pulse duration is showed in FIGURE 14 where:

$Y = I_{DMax}/I_D$ $K = I_{dimp}/I_D$

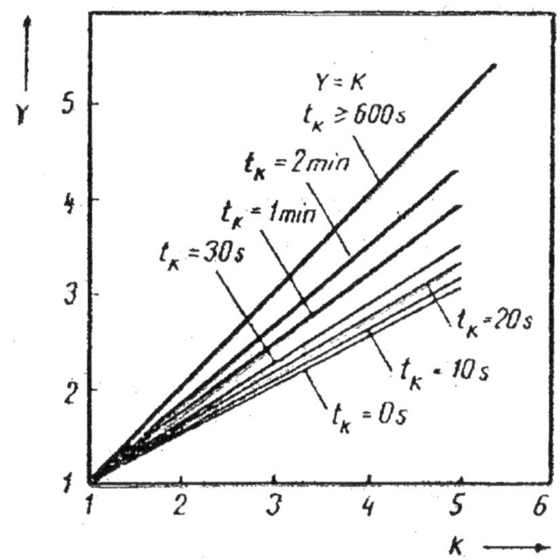

FIGURE 14   Overloads accepted by diodes
function of heat sink size and pulse duration (K and t_K)

In reverse conduction the diode has maximum electrical resistance and the residual current I_R is very small (μA - nA) but very dependent on the temperature.

Increasing the reverse voltage V_R on the diode, the electric field in **pn** junction reaches a critical value when the residual current increases suddenly caused by the multiplication in avalanche of the electric charge carriers. That critical value is called avalanche or breakdown voltage V_{RA}.

There is a special built and tested diode called *diode with control of avalanche* which can support short time high power avalanche impulses in the 200-1000V range. It is marked with the letter *A* and used in rectifiers working on inductive loads or as voltage limiter.

The usual (normal) rectifier diode does not support the avalanche, not even for short time; it can work only in the limits recommended in catalogues and is marked with the letter *N*.

The rapid rectifier diode is marked with the letter *R*.

The rectifying installations usually contain an input transformer to adapt the input voltage of rectifier to its output needed voltage and to decrease the short circuit currents, rectifying elements (diodes, thyristors), ultrarapid protection elements (fuses, interrupters, adaptors), filters.

The rectifying schemas determine the stress of the rectifying elements (diodes and thyristors) at forward currents I_F, I_{FRM}, I_{FSM} and reverse voltages V_R, V_{RRM}, V_{RSM}, the undulation of the rectified voltage, the input transformer power, the output characteristic of the rectifier (voltage - load current) and the effects on the power (the power factor and the distortion regime).

The rectifier realized with diodes only (non-controlled valves) is considered not to be controlled and its output voltage is determined only by the input alternating voltage.

The rectifier is considered to be a controlled one if its output voltage can be modified with controlled valves such as tiratrons, ignitrons and thyristors.

Depending on the number of phases of the input alternating voltage, there are single-phase, three-phase and poly-phase (usually 6 and 12 phases) rectifiers and depending on the number of the rectified alternations, there are half-wave rectifiers and full-wave rectifiers, see FIGURE 15.

The single-phase, half-wave rectifier FIGURE 15a works by stopping the negative half of the waveform using the diode, which conducts only for the positive half of the wave.

The rectified voltage has interruptions and a great number of superior harmonics.

The single-phase, full-wave rectifier with neutral point (middle connection) presented in FIGURE 15b can be considered two half-wave rectifiers in parallel, having the input voltages at 180 degrees phase difference.

For the positive half of the waveform conducts the diode 1 and for the negative half of the waveform conducts the diode 2.

The output voltage is a succession of positive alternations.

The single-phase, bridge rectifier resented in FIGURE 15c can be considered as two middle point rectifiers in series. In series with the load there are two diodes in series, so the power dissipation on rectifier is double. But the reverse voltage on each diode is lower and the transformers are better utilised.

The three-phase rectifier with neutral point or in star connection is presented in FIGURE 15d. Here the output voltage is a direct voltage with three pulses in a period. The bridge three-phase rectifier FIGURE 15e can be considered as two star rectifiers in series at 60 degrees phase difference.

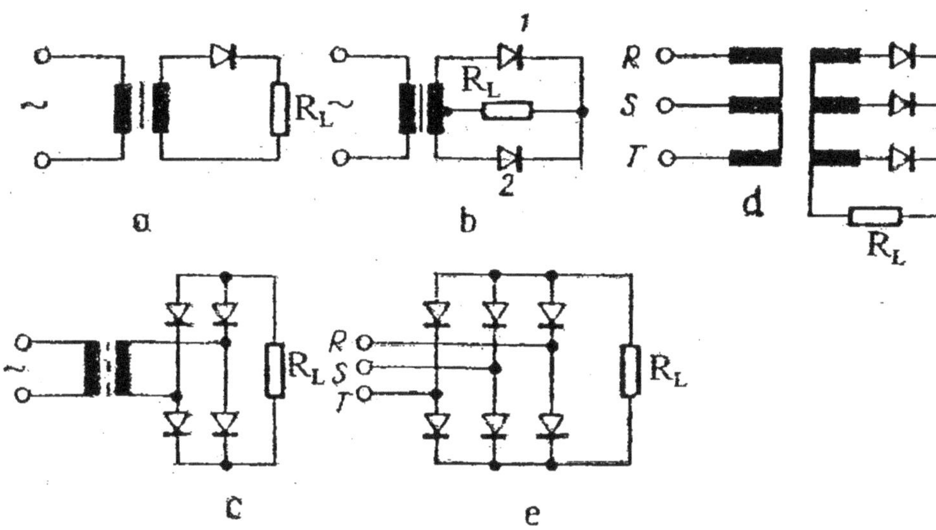

FIGURE 15 Single-phase rectifiers: a. half-wave b. full-wave c. bridge
Three-phase rectifiers: d. star e. bridge

The single-phase, half-wave rectifier is the simplest circuit to convert the alternating current in the direct current using the rectifier diode and is presented in FIGURE 16 together with its output voltage at a sinusoidal input voltage.

When the input voltage is positive, the diode conducts and the voltage at output is equal to the input voltage (neglecting the voltage drop on the diode).

When the input voltage is negative the diode is in reverse conduction and the output voltage is zero. So, the output voltage follows the input voltage for half of period and is zero for the other half.

The direct component of output voltage V_{dc} is the average value of the wave drawn with uninterrupted line in FIGURE 16b and is given by the formula:

$$V_{dc} = [V_m /(2\pi)] \int_0^{\pi} \sin\theta \, d\theta = V_m /\pi$$

V_m is the peak value of the input alternating voltage.

The output direct current I_{dc} through the load resistor R_L is given by the formula:

$$I_{dc} = V_m / (\pi R_L)$$

The rectifier diode should be chosen to support a direct current I_{dc} and a reverse voltage V_m, because when it is off (reverse conduction) the input voltage is applied on it.
The half-wave rectifier gives an output voltage with an alternating component that is comparable with the wanted direct component.

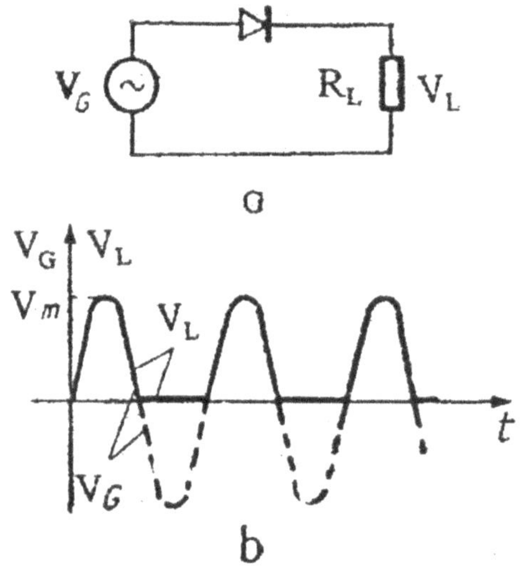

FIGURE 16 Half-wave rectifier a. circuit b. input /output voltage

Another rectifier is the single-phase full-wave rectifier that doubles the direct component of output voltage without increasing the pick of the alternating component and is presented in FIGURE 17.
The transformer in this circuit needs a middle point and $V_2(t) = - V_1(t)$.
When V_1 is positive and V_2 is negative, the diode 1 is in forward conduction and the output voltage follows V_1; when V_1 is negative and V_2 is positive, the diode 2 is in forward conduction and the output voltage follows V_2. The output direct current in the load resistance is twice that of the half-wave rectifier with the same input voltage and the same load R_L and is given by the formula:

$$I_{dc} = 2V_m / (\pi R_L)$$

Each of the rectifier diodes have to stand a direct current $I_{dc}/2$ and a reverse voltage $2V_m$.

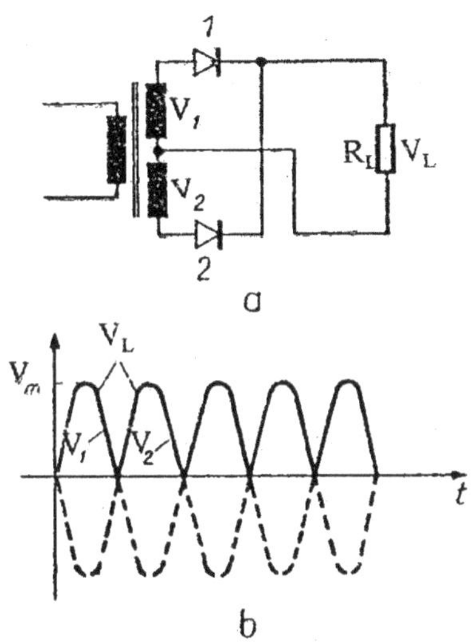

FIGURE 17 Full-wave rectifier with middle point transformer
a. circuit b. input/output voltage

The full-wave rectifier can be realized without transformer if there are used four diodes. The circuit is called the bridge rectifier ad is presented in FIGURE 18.

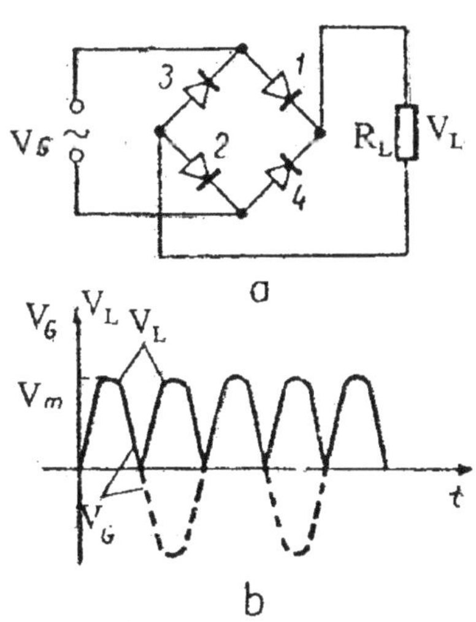

FIGURE 18 Bridge rectifier a. circuit b. input/output voltage

In this circuit when the input voltage V_G is positive the diodes 1, 2 are in forward conduction and the diodes 3, 4 are off. When the input voltage V_G is negative the diodes 3, 4 are in forward conduction and the diodes 1, 2 are off.

The output current in the load resistance R_L is:

$$I_{dc} = 2V_m / (\pi R_L)$$

Each of rectifier diodes have to support a direct current $I_{dc}/2$ and a reverse voltage V_m.

ZENER DIODE

The Zener diode known also as *voltage stabilizer diode* or *breakdown diode* has a well-controlled/non-destructive breakdown voltage because of the Zener effect (discovered by the American physicist Clarence Melvin Zener) or because of the avalanche effect (as in the avalanche diode).

The Zener diode has sharp breakdown characteristics in the reverse bias region and is used in the electronic circuits as stabilized voltage source.

The symbols of the Zener diode are presented in FIGURE 19 and its I-V characteristic curve is presented in FIGURE 20.

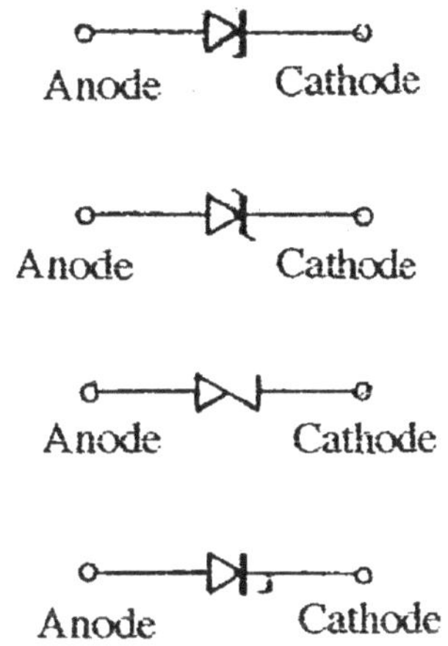

FIGURE 19
Zener diode symbols

For silicon made diodes up to about 5.6V, the Zener effect is predominant exhibiting a marked negative temperature coefficient; above 5.6 V the avalanche effect becomes predominant and exhibits a positive temperature coefficient. In the 5.6V diode the two effects occur together and their temperature coefficients cancel each other out, in consequence the 5.6V diode is a choice for the temperature critical applications.

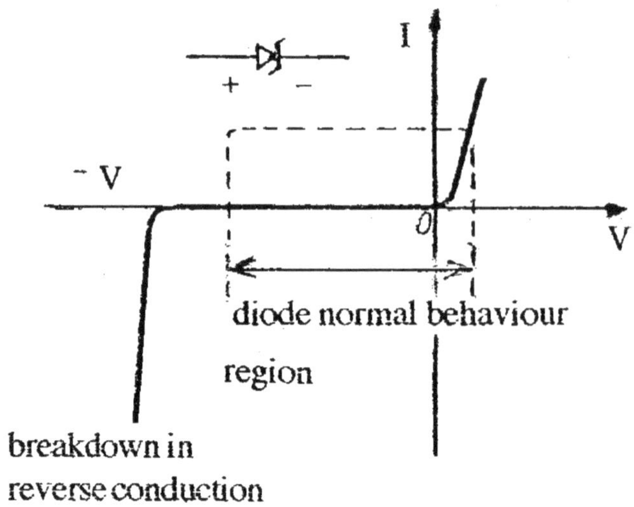

FIGURE 20
Reverse breakdown in a Zener diode

There are Zener diodes with stabilized voltage V_z from 3.5V to 200V, supporting a dissipated power from 250mW to 50W.

The stabilized voltage increases with the semiconductor resistivity (usually silicon).

The parameters that characterize the Zener diode are:

V_z nominal Zener voltage (V)

I_z control electric current of the voltage V_z

r_z dynamic electrical resistance in the stabilizing region calculated at a specified current I_z $r_z = \Delta V_z / \Delta I_z$ (Ω)

å$_{VzT}$ temperature coefficient of the stabilized voltage with values 0.07...0.37 %/°C
å$_{VzT}$ $= (\Delta V_z / \Delta T)(1/V_z)$ (% / °C or 10^{-4}/°C)

r_z and å$_{VzT}$ should have low values

Q quality coefficient $Q = (\Delta V_z / V_z) / (\Delta I_z / I_z)$

I_{zM} maximum Zener current (A)

T operating temperature (°C)

P_{Dmax} maximum dissipated electric power (W)

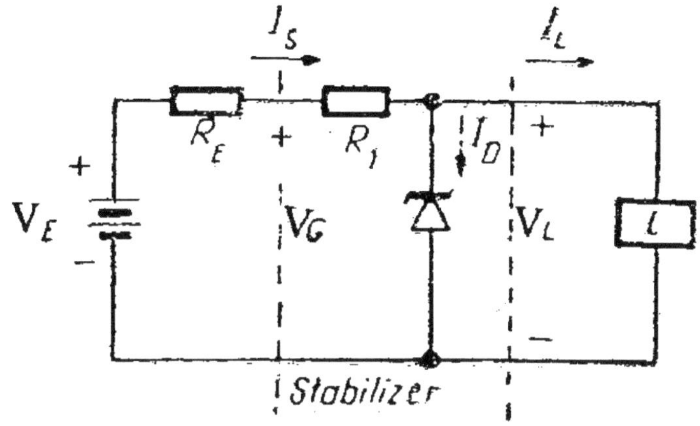

FIGURE 21 Zener diode used as voltage stabilizer

The Zener diodes are utilised as voltage stabilizers for voltage suppliers.

It is known that the output voltage of a rectifier used as voltage supplier decreases when the load current increases. That dependence can be reduced by using a Zener diode.

In the circuit from FIGURE 21 the voltage supplier is represented by its Thevenin equivalent circuit containing V_E and R_E.

The stabilizer is represented by the Zener diode and the serial resistor R_l.

The Zener diode is also used as voltage limiter, signal maker and in protection circuits.

COMMUTATION DIODE/IMPULSE DIODE

The commutation diode known also as *impulse diode* is the diode specially manufactured to work in the commutation regime ON-OFF/OFF-ON.

Its electrical capacitance is made very low.

The I-V characteristic curve of the commutation diode is not different from the general I-V characteristic curve of the diode.

In catalogues for the commutation diode besides the I-V characteristic curve is given the graphical variation of the **pn** junction capacitance with the reverse voltage.

The diode in the commutation regime is characterized by:

static value of its electrical resistance	(Ω)	in forward conduction should be low
		in reverse conduction should be high
parameters of the transition process	(s)	forward commutation time (OFF-ON)
		reverse commutation time (ON-OFF)

The forward commutation time t_f is 1ns - 20 ns.

The work speed of the commutation circuits with diodes is limited by the reverse commutation time t_r, actually reverse recovery time t_{rr} , which is relatively long: 50μs for usual diodes to 1μs for special gold doped diodes. That is composed by the reverse elimination time t_{re} to eliminate mobile electric charge carriers in minority (with the life time t_m) accumulated in junction and the reverse falling time t_{rf} comparable with forward commutation time t_f and much smaller than t_r, see FIGURE 22.

Trying to minimise t_{rr} to have rapid commutation there are used two methods:

diminish the life time of mobile electricity carriers by increasing the number of recombination centres(although will increase the reverse current too)

utilise short life time mobile electricity carriers semiconductors (GaAs)

The experiment did not find connections between the static parameters and the commutation times of the diodes. The behaviour of the diode in the transitory regime is determined correctly by practical measurements.

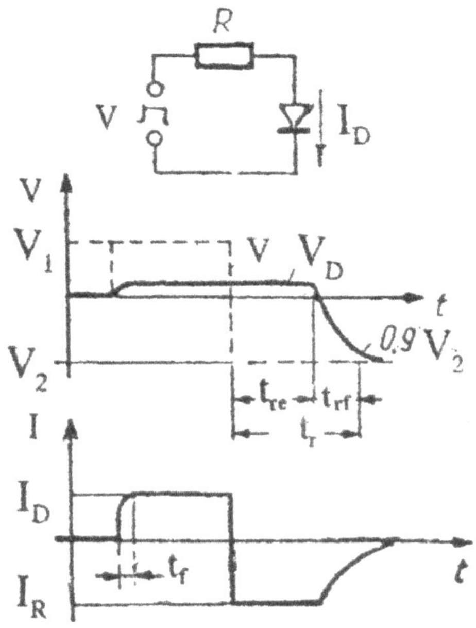

FIGURE 22

Voltage and current variation at the diode terminals in the commutation process

$t_{rr} = t_{re} + t_{rf}$ $\quad$ $t_{re} = t_m \ln(1+I_D/I_R)$ $\quad$ $I_D = (V_1 - V_D)/R$ $\quad$ $I_R = (V_2 - V_D)/R$

The most important parameters in connection with the commutation regime are:

junction capacitance at zero bias

junction capacitance in reverse conduction

reverse commutation time

electric charge accumulated in forward conduction

It is possible to realize very short commutation times if the diode is manufactured with metal-semiconductor junction (behaving similar to **pn** junction); that diode, called *Schottky diode* has fast switching times (<1ns) and also low forward voltage drop (0.15 – 0.45 V) making it useful in very high frequency circuits (GHz) and in high efficiency rectifiers.

For high frequency it is used also a diode with different structure at which between the **p** and **n** regions of the **pn** junction is introduced an intrinsic material layer.
That is the *PIN diode*.
The "invading" of the intrinsic region by the mobile electric charge carriers and their long life in this region, make the diode to lose its rectifying property at frequencies above 10-20 MHz. At those frequencies, the diode behaviour is that of a resistance with the value controlled by its direct current.
The above-mentioned features permit the use of the PIN diode in the commutation of the high frequency signals or as electronic controlled resistance.
FIGURE 23 shows the use of the impulse diodes in logic circuits OR (a) and AND (b).
The logic circuits are used in numerical electronic devices as the electronic calculator.

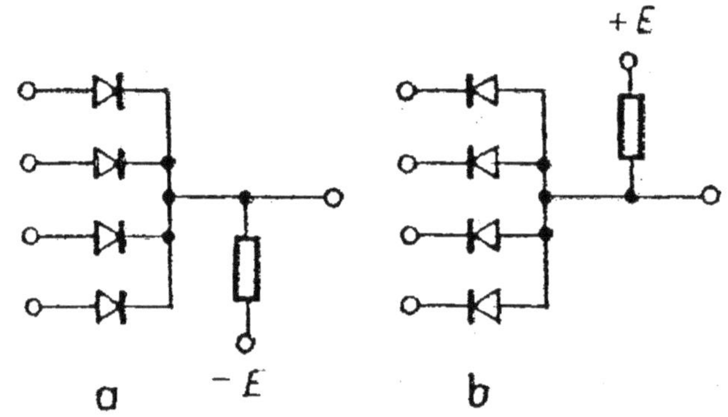

FIGURE 23 Logic circuits with diodes a. OR circuit b. AND circuit

The commutation diode is switched ON-OFF or OFF-ON by the "processed" signal itself.
There is also the *diode for electronic commutation* which is switched ON or OFF by a direct voltage permitting or not the alternating signal to pass.
In FIGURE 24 is given an example with diodes for electronic commutation used in the circuit for the permutation of lines in the colour television.

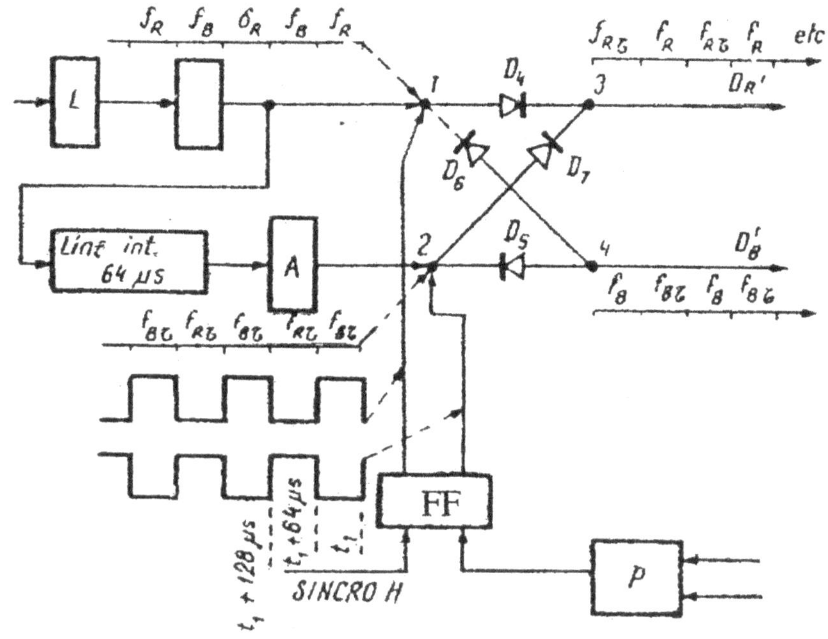

FIGURE 24 Electronic commutator/switch with diodes for colour TV

DETECTION DIODE

The detection diode works in the range of the metre, decimetre and centimetre wavelength electromagnetic signals to convert their frequencies (mixing diode) and/or to demodulate the radio/video signals.

The detection is alike to rectifying but the "processed" signal has frequencies in the range of hundreds kHz/MHz not 50Hz-20kHz.

The electric power dissipation on this diode is low (not significant).

Because of the frequency range in the manufacturing of the detection diode are adopted solutions specific for high frequencies to realize structures with surfaces as small as possible to:

have low parasite capacitances and inductances

facilitate the insertion in circuits (resonant cavities, co-ax feeders etc)

The equivalent circuit of the detection diode, actually a **pn** junction is presented in FIGURE 25.

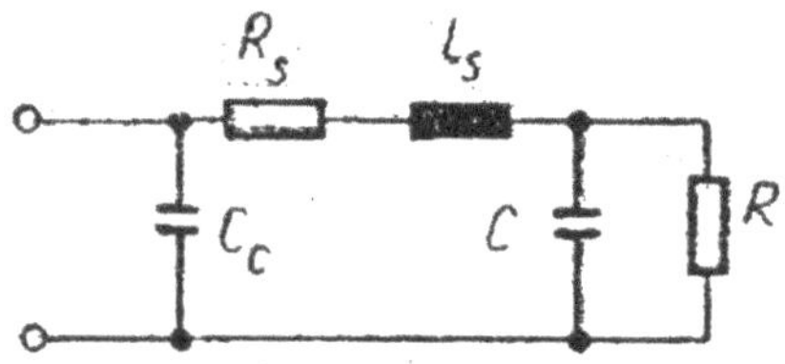

FIGURE 25 Equivalent circuit of detection diode

R electrical resistance of the junction

C capacitance of the junction

R_s serial electrical resistance of the junction

L_s serial inductance of the junction

C_c intrinsic capacitance of the diode

The I-V characteristic curve of the detection diode is not different from that of the rectifier diode.

In FIGURE 26 is presented the electronic circuit of the TV video detector.
The diode EFD-106 is a detection diode for the video signal.

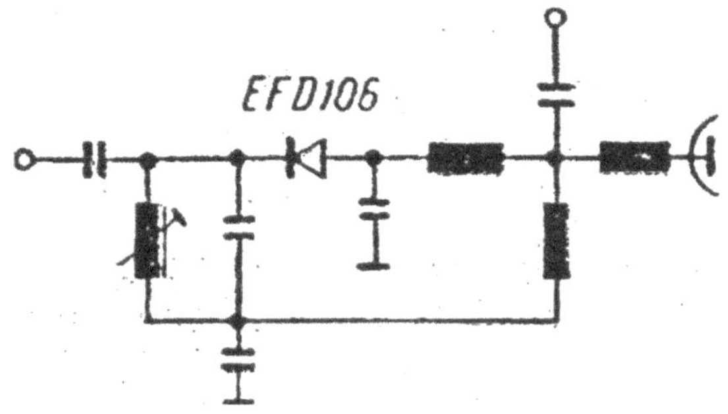

FIGURE 26 Video detector circuit used in *Electra* TV

VARICAP DIODE

The varicap diode known also as *varactor diode* is the diode that in reverse conduction presents an electrical capacitance, function of the applied reverse voltage V_R.
The word "varicap" comes from ***vari****able* ***cap****acitance.*
The word "varactor" comes from ***var****iable reactance/re****actor****.*
The symbols of the varicap diode are presented in FIGURE 27.

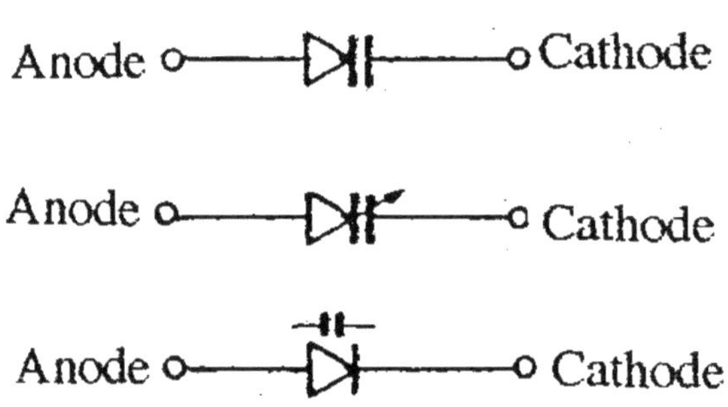

FIGURE 27 Varicap diode symbols

The electrical capacitance of the diode is varied comfortable varying the reverse voltage V_R.
It provides an electrically controllable capacitance, which can be used in tuned circuits.
The equivalent circuit of the varicap diode, actually a **pn** junction operated reverse-biased is presented in FIGURE 28 where:

C variable capacitance of the **pn** junction
R_s equivalent serial electrical resistance
L_s equivalent serial inductance
C_c intrinsic capacitance of the diode

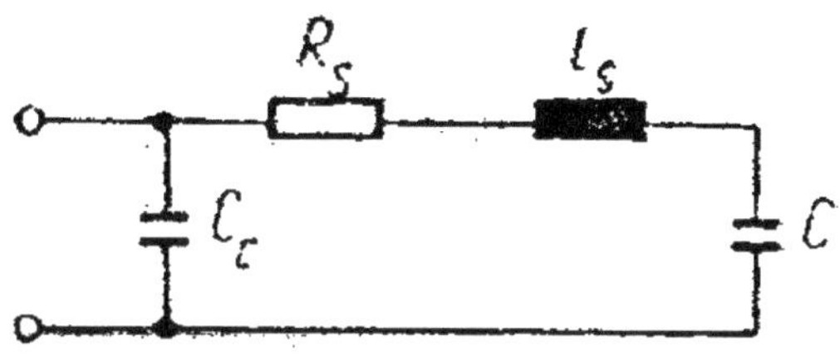

FIGURE 28 Varicap diode equivalent circuit

The diode capacitance decreases as the reverse bias increases, according to the relation:

$$C = C_o / (1+V_R / V_o)^n$$

C_o and V_o are constants. V_o is approximately the forward voltage of the diode. The exponent *n* depends on how the doping density depend on distance away from the junction.

n=0.33 for a graded junction (linear variation).

n=0.5 for an abrupt junction (constant doping density).

n=2 for a hyper-abrupt junction (density jumps abruptly at the junction, then decreases) obtaining a maximum variation of the capacitance for a given variation of the reverse voltage.

The specific parameters of the varicap diode are:

V_{R1}-V_{R2} voltage range for which the diode has a variable capacitance

C_{max} **pn** junction capacitance C at null voltage

C_{min} **pn** junction capacitance C at maximum reverse voltage

C_c diode intrinsic capacitance

L_s equivalent serial inductance

R_s equivalent serial electrical resistance

f_c *cutting frequency* at which the quality factor of the diode becomes equal with unity, the capacitance having a minimum value corresponding to the maximum reverse voltage

$f_c = 1/(2\pi R_s C_{min})$ (Hz)

In FIGURE 29 is presented the behaviour of the varicap diode BA102.

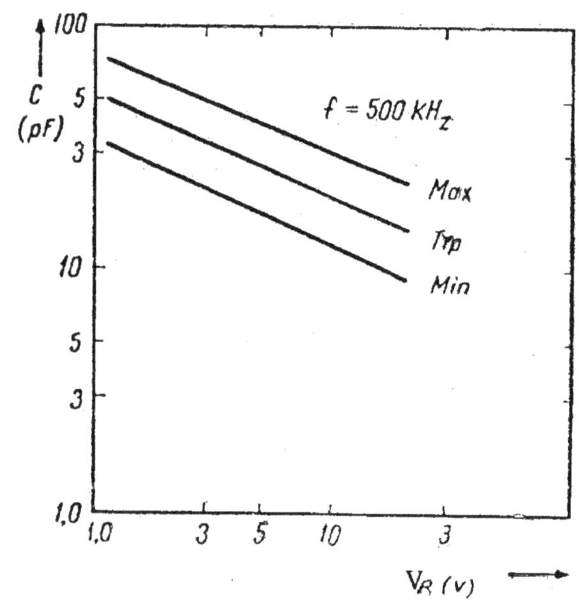

FIGURE 29 Diode capacitance / Reverse voltage for the diode BA 102

It is important to select a diode that combines the correct electrical capacitance range for the available tuning range. The higher voltage specified is normally the maximum reverse bias and this should not be exceeded because may occur the breakdown.

The varicap diode is used as voltage-controlled capacitance in parametric amplifiers, parametric oscillators and voltage-controlled oscillators, frequency synthesisers, filters. Built with hyper-abrupt junction and named *tuning diode* is used to tune without commutation in radio receptors and TVs.

TUNNEL DIODE/ESAKI DIODE

Discovered by the Japanese physicist Reona/Leo Esaki in the year 1958 the tunnel diode is often called *Esaki diode.* Its main attractiveness is the high speed capability useful for microwave modulation, mixing and detection.

Tunnel diode symbols are presented in FIGURE 30.

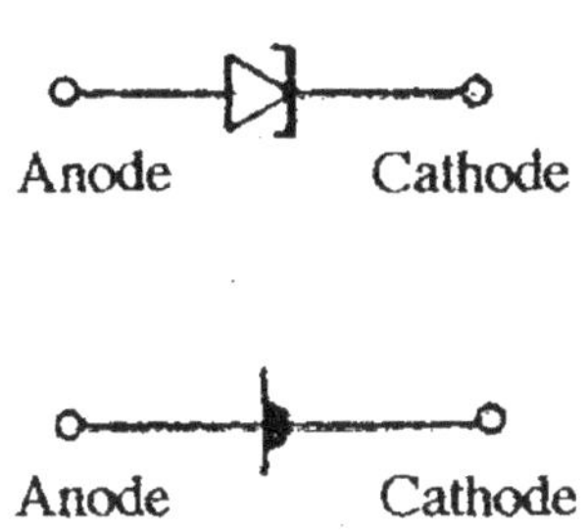

FIGURE 30 Tunnel diode symbols

When a **pn** junction is very strong contaminated in both regions becoming **p+n+** (more than 10^{19} impurity atoms / cubic centimetre) the Zener effect can appear at very low reverse voltages, zero voltage inclusive.

A diode with the Zener effect at zero voltage can support high reverse currents and returns normal for positive voltage.

0.1V - 0.2V forward voltage is sufficient to eliminate the Zener mechanism and reduce the current through the junction. Further increasing of the forward voltage will result in the current increasing.

The I-V characteristic curve of the above description is presented in FIGURE 31.

The device having that kind of behaviour is the tunnel diode; the name comes from the explanation given in the quantum mechanics for the Zener breakdown effect.

The tunnel diode is useful too because has a negative differential electrical resistance region shown where the I-V characteristic curve has a negative slope (N shaped).

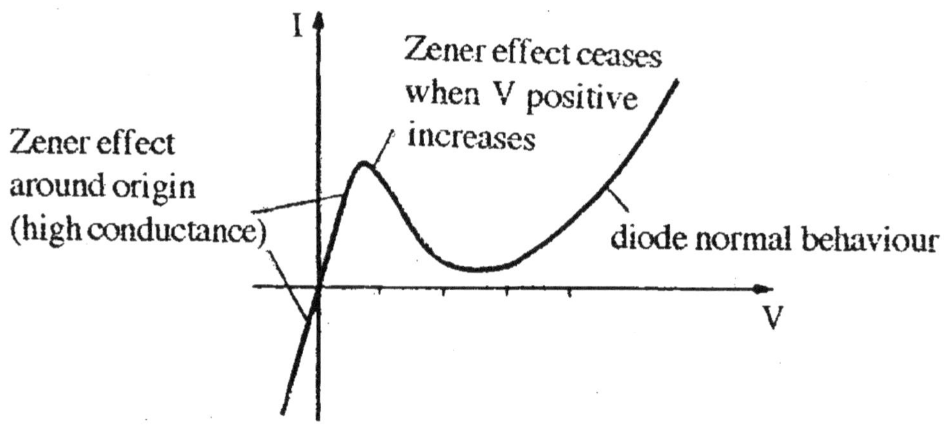

FIGURE 31 I-V characteristic of the tunnel diode

The negative differential resistance can compensate the losses in a resonant circuit permitting the generation of very high frequency oscillations up to 10 GHz at very low power (mW).
The negative differential resistance appears in a large temperature range, so for example for the germanium manufactured diodes was observed that the negative differential resistance is maintained in the temperature range –265° C … +250° C and for the silicon manufactured diodes the upper limit increases to 350° C.
Briefly the tunnel diode is an one direction, low power, rapid, with negative differential electrical resistance semiconductor device. It works in a large temperature range and is not influenced by the nuclear radiation and since is manufactured with strong contaminated semiconductors it is not necessary to use very pure materials.
It is mostly confined to discrete components since its incompatibility with the integrated-circuit technology.

The four parameters of the tunnel diode are:

voltage and current of the "pick" of the I-V characteristic curve

voltage and current of the "valley" of the I-V characteristic curve

negative differential electrical resistance $R = \Delta V/\Delta I = \text{ctg } å$

voltage trip (voltage between the "pick" of the I-V characteristic curve and the point corresponding to same current on characteristic in normal behaviour)

With the tunnel diode it is possible to realize circuits that work at hundreds and thousands of GHz and permit commutation times in the nanosecond range.

Mostly is used as microwave low noise amplifier.

In FIGURE 32 is presented the logic circuit AND realised with a tunnel diode.

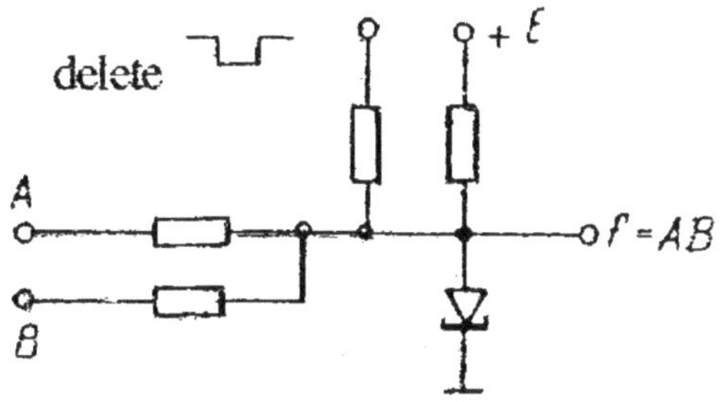

FIGURE 32 AND logic circuit with tunnel diode

Below are given the parameters of the germanium tunnel diode TR9A:

Voltage of "pick"	45	mV
Current of "pick"	1	mA
Voltage of "valley"	170	mV
Current of "valley"	0.25	mA
Voltage trip	285	mV
Negative conductance	-9	mho
Junction capacitance	25	pF

IMPATT DIODE

The IMPATT diode is the most powerful solid-state microwave generator in the range 3-300 GHz.

The word *IMPATT* comes from ***imp**act-ionisation-**a**valanche **t**ransit-**t**ime.*

The IMPATT diode works in the avalanche breakdown regime and its function is based on the fact that has a dynamic negative electrical resistance in the GHz frequency range.

The static characteristics of an IMPATT diode are not different from a standard **pn** junction diode. These characteristics do not show conventional negative differential electrical resistance (negative dV/dt). The negative resistance of an IMPATT diode is dynamic in nature such that when a sinusoidal voltage is applied to an IMPATT diode the current responds with 180° phase lag.

The equivalent circuit of the IMPATT diode presented in FIGURE 33 has a resistance r that includes the dynamic negative resistance and the parasitic positive resistances in serial connection with the reactance X (with capacitance character determined first by the **pn** junction capacitance at the breakdown voltage).

FIGURE 33 Equivalent microwave circuit of the IMPATT diode

The negative value of the resistance r in a certain range of frequencies (the range of frequencies depends on the constructive features of the diode) permits the use of the IMPATT diodes in circuits which generate or amplify the high frequency signals. In the first case the oscillation frequency is determined by the oscillator formed with the diode capacitance and the load inductance (including the parasitic inductances of the capsule, connections, etc).

To bias the IMPATT diode in circuit is recommended a constant current generator.

The bias current should be increased gradually to the operating value.

The generated power P_{out} increases with the biasing current, reaching a maximum.

Another diode very useful as microwave device in the range 1-100GHz is the *Gunn diode* known also as ***t**ransferred-**e**lectron **d**evice TED.*

TED has a voltage-controlled negative differential electrical resistance (N shaped).

It is used mainly for microwave generation as ***t**ransferred-**e**lectron **o**scillator* TEO and for microwave amplification as ***t**ransferred-**e**lectron **a**mplifier* TEA.

DOUBLE-BASE DIODE/UNIJUNCTION TRANSISTOR

The diode known in 1950s as the double-base diode was known in 1949 year as the *filamentary transistor* and from 1960s as the ***unijunction transistor*** UJT.
The double-base diode has three electrodes and is an active device in oscillation and commutation circuits.
As ordinary structure the double-base diode is an uniform contaminated mostly **n** silicon bar having an ohmic contact at each end; in the middle is partially **p** contaminated so has a **pn** junction. Complementary devices on **p** silicon bar are also possible.
The two ohmic contacts at the ends are called the base1 B_1 and the base2 B_2 and the ohmic contact to the **p** region is called the emitter E.
The symbol of the double-base diode is presented in FIGURE 34.

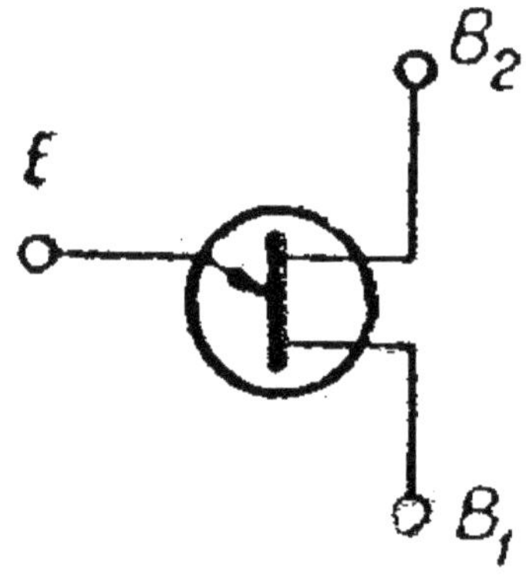

FIGURE 34
Double-base diode symbol

The structure and circuit connection of the double-base diode is presented in FIGURE 35.

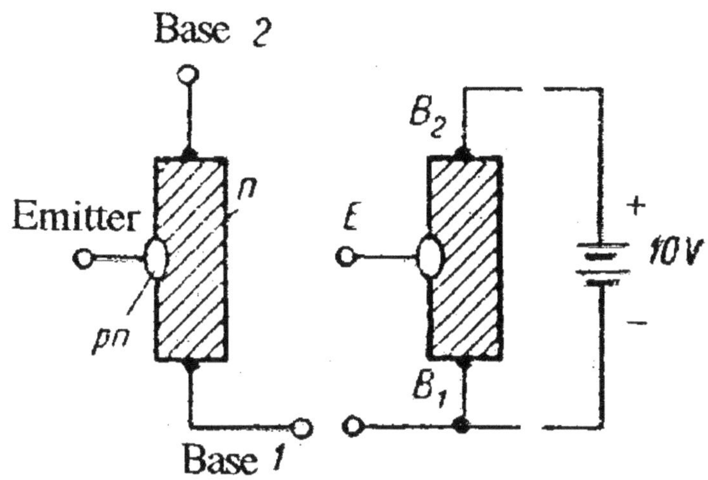

FIGURE 35 Double-base diode

The electrical resistance of the silicon bar is in the range 4 – 10 k Ω.
If the two bases are connected together (V_{B2B1} = 0V) the device is a normal diode.
The UJT behaviour begins usually when V_{B2B1} = 10V.
If the emitter is located just in the middle of the bar it will be biased at V_1=+5V from Base1.

When the emitter-base1 voltage V_{EB1} is less than V_1 the **pn** junction is in reverse conduction, the emitter-base1 current I_{EB1} is small and the diode is in the cut-off regime in the OA region of the I_E-V_{EB1} characteristic curve FIGURE 36a.

When the voltage applied between emitter and base1 V_{EB1} is higher than the value V_1, the **pn** junction is in forward conduction and the emitter-base1 current I_{EB1} will increase, how shows the characteristic curve which presents a negative resistance too; the emitter-base1 circuit injects current into the base2-base1 circuit, the bases circuit conducts and a negative impulse appears at the B2 terminal (V_{OB2}) and a positive impulse appears at the B1 terminal (V_{OB1}).

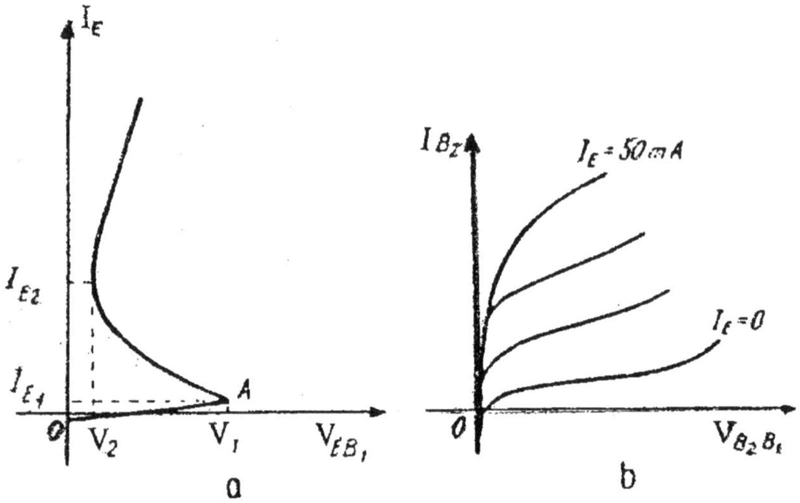

FIGURE 36 I-V characteristics of UJT

From the point where the voltage V_{EB1} has the minimum value V_2 follows the normal forward characteristic curve of a diode; this region of the characteristic is called the saturation region and that is better presented by the family of characteristic curves I_{B2}-V_{B2B1} for different emitter currents I_E from FIGURE 36b.

The electrical parameters of the double-base diode are:

V_{B2B1}	voltage between the bases
R_{B2B1}	electrical resistance between the bases
$n=R_{EB1}/R_{B2B1}$	intrinsic ratio
V_1	maximum voltage V_{EB1}
V_2	minimum voltage V_{EB1}
I_{E1}	emitter electric current for maximum V_{EB1}
I_{E2}	emitter current for minimum V_{EB1}
I_{EBO}	emitter reverse current for $V_{EB1} < (1/2)V_{B2B1}$
I_E*	impulse emitter current
V_{OB1}	peak pulse voltage at B_1

The forward and reverse commutation times of UJTs are in the microseconds range.

The UJTs are used as active devices for sawtooth waveform oscillators and multivibrators or in the commutation of the thyristors and triacs.

In FIGURE 37 is presented a controlled rectifier whose triac is switched ON with the help of UJT connected in a conventional circuit for phase control.

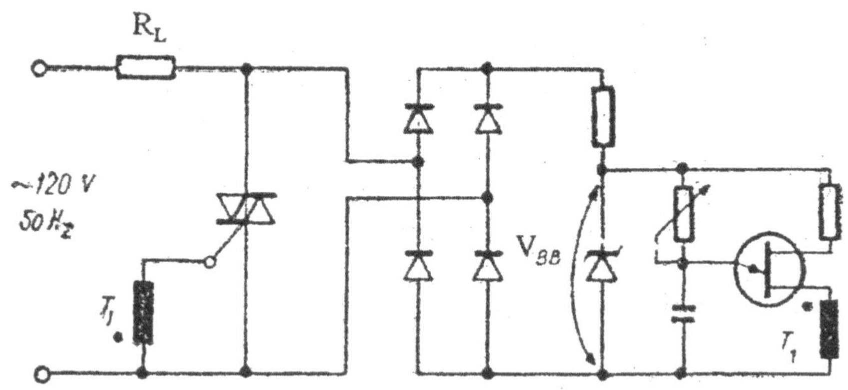

FIGURE 37 Manually controlled rectifier using UJT

CONTROLLED DIODE/THYRISTOR

The controlled diode, better known by the name *thyristor* is a **pnpn** semiconductor structure (3 **pn** junctions) with an electrode for control; it works stable only in two states ON and OFF and is a typical commutation device. Besides the two electrodes of an ordinary diode, anode A and cathode C, the controlled diode has the third electrode for its own control called *gate G*.

The structure and the symbol of the controlled diode are presented in FIGURE 38.

The electric current in OFF state is the reverse conduction current of the **pn** junction and in ON state the controlled diode permits a higher electric current, a consequence of the avalanche multiplication process in the middle junction.

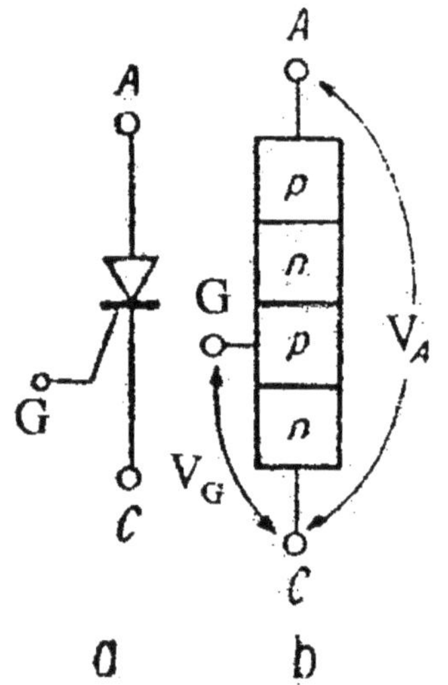

FIGURE 38 Thyristor a. symbol b. structure

The behaviour of the diode without the control voltage on the gate is presented in FIGURE 39.

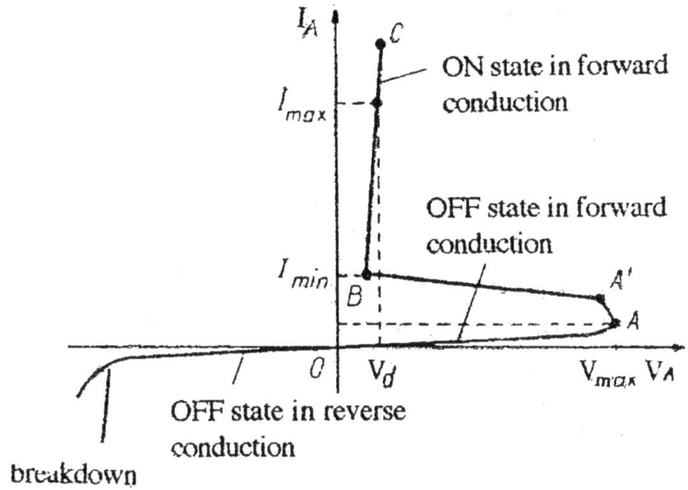

FIGURE 39 I-V characteristic for a thyristor without gate control

The diode is switched ON at a voltage between anode and cathode $V_A = V_{max}$ and for lower voltages it is in the OFF state. The electric current is limited by external resistor R.

Features: anode-cathode electrical resistance in OFF state is in the range $10^8 - 10^9\ \Omega$
anode-cathode electrical resistance in ON state is in the ohms range
anode-cathode voltage in ON state is approximately 2V

If between the gate and the cathode is applied a positive voltage the thyristor switches ON for anode-cathode voltages lower than V_{max} FIGURE 40.
The forward commutation time (OFF-ON time) t_g is minimum of the order 10^{-8} seconds.
The reverse commutation time (ON-OFF time) t_q is the while between the moment when the forward current is zero (as consequence of changing the anode-cathode circuit) and the moment when the thyristor can support again the anode-cathode voltage without to open and is minimum of the order 10^{-6} seconds.

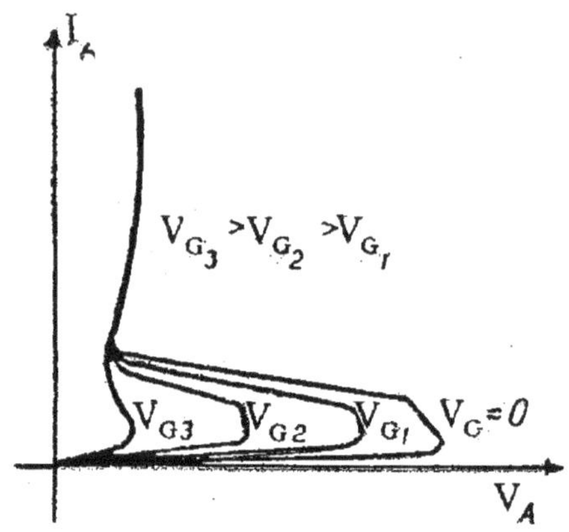

FIGURE 40 I-V characteristics of the controlled thyristor

The gate can control only the switching in the forward conduction OFF-ON.
The switching in the reverse conduction ON-OFF is done by nullifying or reversing the anode-cathode voltage or decreasing the current under the self-maintaining value.

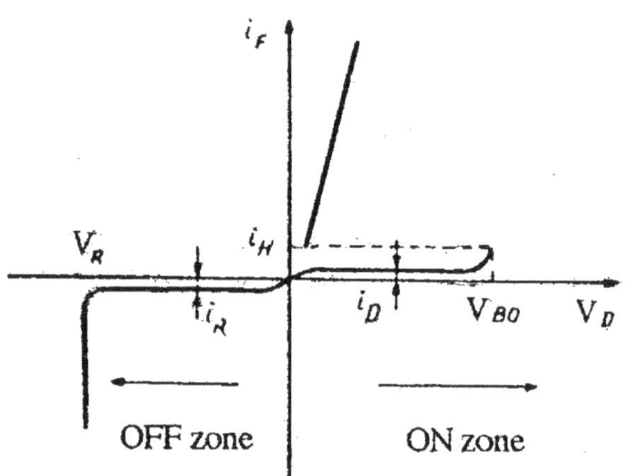

FIGURE 41 Stylised characteristic of a thyristor

Using the ideal characteristic curve from FIGURE 41 it is possible to define the following parameters of the thyristors specified in catalogues:

Forward conduction

I_D forward direct electric current in OFF state at forward bias (A)

V_B positive anode-cathode voltage for which at a command on gate the thyristor enters ON state (V)

V_{B0} V_B for zero gate current

I_H static maintaining current, minimum forward direct current to keep the thyristor ON

I_{HD} dynamic maintaining current, the minimum forward current that should appear at OFF-ON switching moment ; this value depends on the amplitude, shape and duration of the command impulse and is higher than I_H

i_F instantaneous value of the current in ON state

I_F average value for a period for i_F

I_{FRM} maximum repetitive I_F

I_{FSM} maximum surge (peak) I_F

I_N nominal current of thyristor, recommended by manufacturer for I_F for monophase circuit and resistor as load

This value depends on the type of heat sink (radiator) and ventilation

I_{FL} limit current I_F for long time work, in a monophase schema and resistor as load

$I^2 t$ represents the maximum permitted value of the integral in time of the thyristor current's square calculated for 10ms and 1ms; it is useful for sizing the protection elements of the thyristors (fuses, switches, automatons)

Reverse conduction

I_R reverse current through the thyristor in OFF state

V_{FRM}, V_{RRM} maximum repetitive voltage, positive and negative in OFF state

V_{FSM}, V_{RSM} maximum surge (peak) voltage, positive and negative in OFF state

Command/Control

I_G command current that passes the control (command) electrode called gate

V_G command voltage on gate

Dynamic regime

t_g forward switch time

t_q reverse switch time

$(dv/dt)_{max}$ maximum speed of increasing the forward voltage

Over this value the thyristor can switch ON without a command impulse

$(di_F/dt)_{max}$ maximum speed of increasing the current

Over this value in the central zone appears additional power dissipation that overheats and destroys the thyristor

Thermal regime

T_j junction temperature, over its maximum value the thyristor is destroyed

T_c capsule temperature

T_A cooling environment temperature, I_N is given for this temperature

P_D power dissipation, over its maximum value the thyristor is destroyed

$R_{thJC} = (T_j - T_c)/P_D$ internal thermal resistance

$R_{thCA} = (T_c - T_A)/P_D$ external thermal resistance

$R_{thJA} = (T_j - T_A)/P_D$ total thermal resistance

The thyristors are used in the controlled rectifiers having characteristics alike with the thyratrons.

In the rectifiers where the voltages and the currents go beyond the limits of one thyristor there is used the series and parallel connection of more thyristors (as for diodes).

The dissipation and super-loads problems are similar to those for the silicon rectifier diodes.

In the rectifiers that use as rectifying elements thyristors, those being controlled valves, the output voltage can be decreased continuously to zero, delaying the command signal on gate.

In FIGURE 42 a,b,c,d,f are presented single-phase and three-phase full controlled rectifiers.

In FIGURE 42 e,g are presented single-phase and three-phase semi-controlled rectifiers.

The function of these schemas is alike to those presented for the rectifier diode, but here the half-waves can be cut function of the command angle.

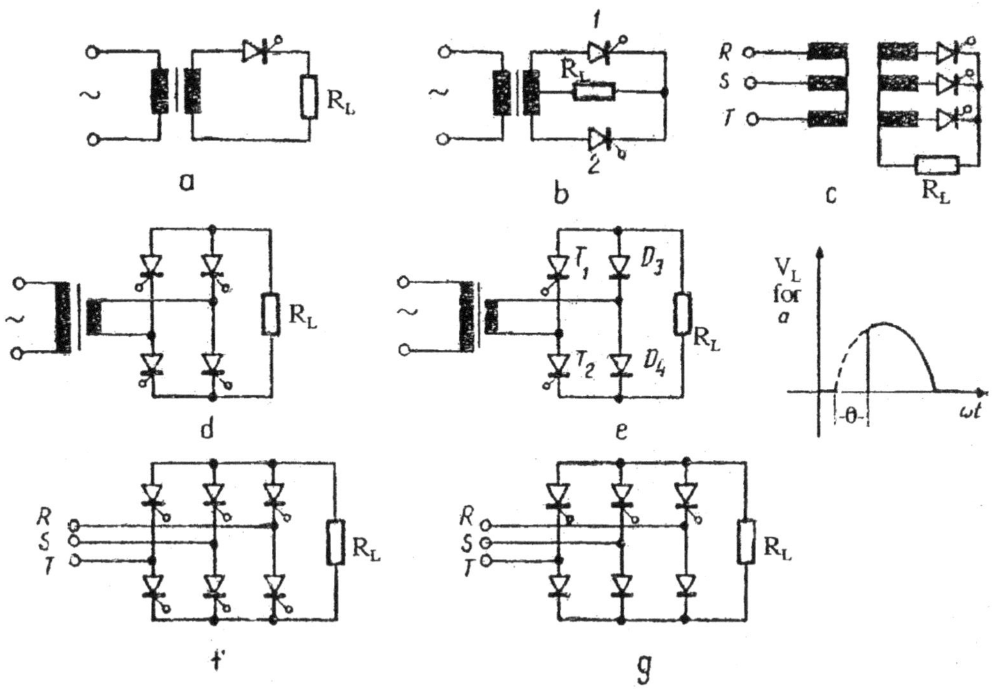

FIGURE 42 Single-phase (a b d e) and three-phase (c f g) rectifiers with thyristors

The average value of the rectified voltage in the full controlled circuits is:

$$V_{med} = (\sqrt{2}m/\pi)\, V \sin(\pi/m) \cos\theta$$

m phase number (single-phase bridge m=2; three-phase star m=3; three-phase bridge m=6)

V effective value of the phase voltage

θ control angle, considered from the moment of the natural entering in conduction of the not controlled valve

The average value of the rectified voltage in semi-controlled bridges is:

$$V_{med} = (\sqrt{2}m/\pi)\, V \sin(\pi/m)\, (1+\cos\theta)/2$$

m phase number (single-phase semi-contr.bridge m=2; three-phase semi-contr.bridge m=6)

There is a thyristor-type device that can conduct in both directions following the positive or the negative command signal on the gate. It is called the *triac*.

The symbol and the basic structure of the triac are presented in FIGURE 43.

Its I-V characteristic curve is presented in FIGURE 44.

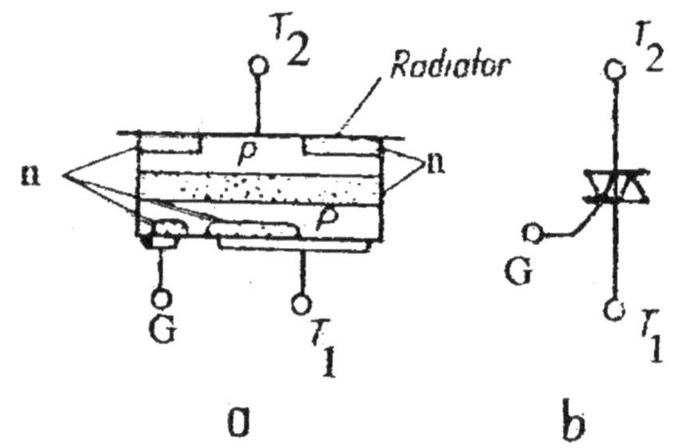

FIGURE 43 Triac a. structure b. symbol

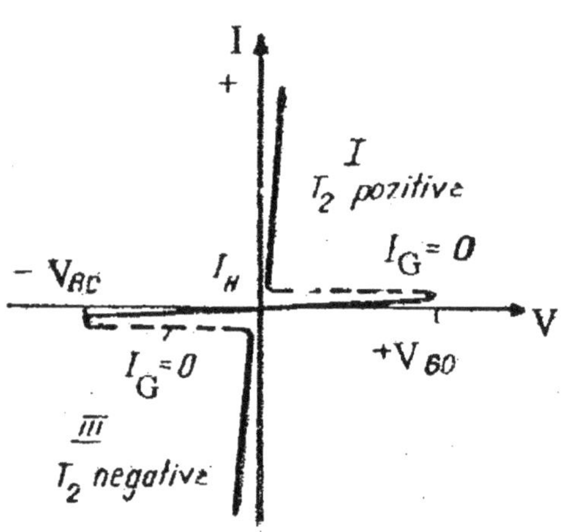

FIGURE 44 Triac I-V characteristic

Other thyristor-type devices are the *Silicon Unilateral Switch*, the *Silicon Bilateral Switch*, the *Dhynistor* and the *Diac*.

Silicon Unilateral Switch SUS is a miniature controlled diode with an anode gate (instead a cathode gate) and a Zener diode connected between gate and cathode.
It is less temperature dependent and requires a lower voltage to trigger than the thyristor. Generally is used in low power applications. Its symbol, its equivalent circuit and its I-V characteristic curve are presented in FIGURE 45.

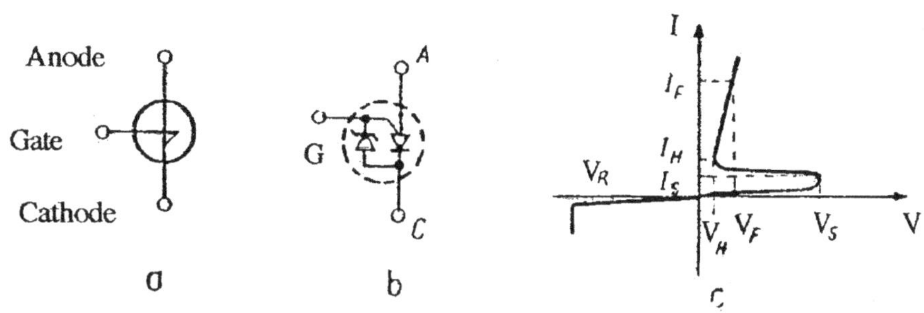

FIGURE 45 SUS a. symbol b. equivalent circuit c. I-V characteristic

Silicon Bilateral Switch SBS is a structure formed with two devices SUS in antiparallel connection. Since SBS works as a switch for both positive and negative voltages it is used to open the triacs. Its symbol, its equivalent circuit and its I-V characteristic curve are presented in FIGURE 46.

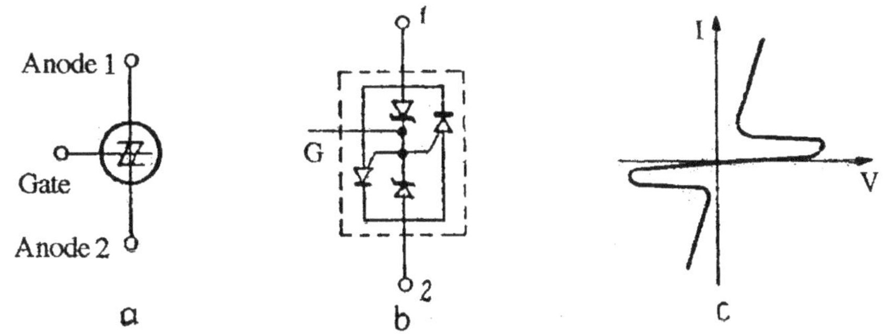

FIGURE 46 SBS a. symbol b. equivalent circuit c. I-V characteristic

Dhynistor or *four-layer diode* has the **pnpn** structure of the thyristor but only two electrodes, anode & cathode, and switches ON for a certain value of the forward voltage. It is used in the control circuits for the thyristors and the triacs.
Its I-V characteristic and its use to control a thyristor are presented in FIGURE 47.

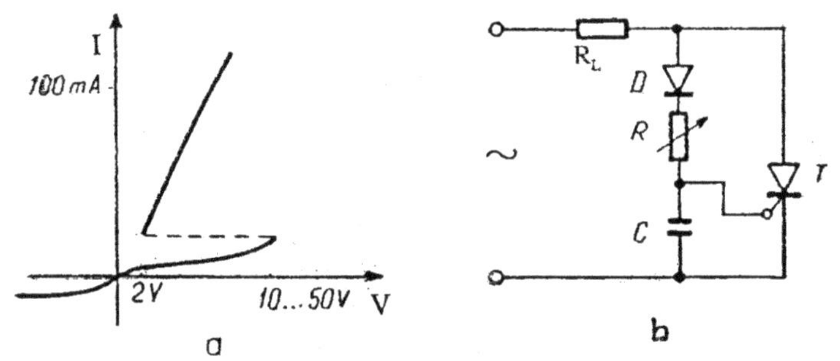

FIGURE 47 Dhynistor a. I-V characteristic b. thyristor controlled by dhynistor

Diac or *bilateral trigger diode* is the electronic semiconductor device with two electrodes which opens when the voltage applied between the electrodes is higher than a certain value V_{B0} (tens volts) equal of its polarity. The diac is used in relaxing oscillators and in switch circuits to command the thyristors and the triacs.
In FIGURE 48 are presented the symbol of the diac, the I-V characteristic curve of the diac with the negative resistance region and a switch schema with diac for a triac used as an alternating current connection, the triac being triggered ON at the moment determined by the RC time constant of the switch circuit.

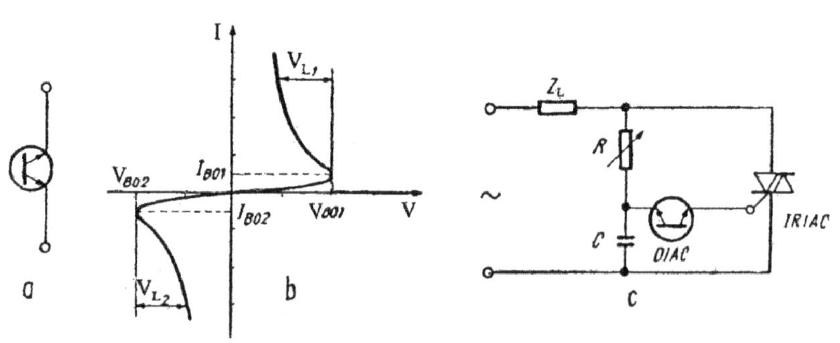

FIGURE 48 Diac a. symbol b. I-V characteristic c. diac controlling a triac

PHOTODIODE

The photodiode is an optoelectronic device that transforms the electromagnetic radiation of the light in the electric energy.
Its function is based on the *photoelectric effect* discovered by the German physicist Heinrich Rudolf Hertz.

The photoelectric effect *or* Hertz effect *is the phenomenon related to the emission of electrons from matter upon the absorption of electromagnetic radiation such as light, ultraviolet radiation or X-rays.*
For practical purposes are taken in consideration:

The external photoelectric effect *is the emitting of the electrons by a solid when it is lit (photoelectric emission).*
There is a minimal energy of the light quanta (photons) hf_0 *to be able to extract the electrons from metal or semiconductor surfaces dependent on the material.*
If the energy of photons is higher, the emitted electrons (photoelectrons) can move with the speed:

$$v = \sqrt{2(hf - hf_0)/m}$$

h = *Planck's constant* $6.626068 \cdot 10^{-34}$ $m^2 \cdot Kg \cdot s^{-1}$
f_0 = *frequency corresponding to the minimal energy of the photon (extraction work)*
f = *frequency corresponding to the energy of the photon*
m = *electron mass*
v = *speed of the emitted electron*

The external photoelectric effect is applied in the manufacturing of the photoelectric cell, which is composed of a glass bulb, enclosing vacuum or inert gas, with two electrodes inside connected at a direct voltage. There is an electric current between electrodes only if the cathode is lit.

The internal photoelectric effect *is the liberation of the electrons inside a semiconductor by the action of the light.*

The photoelectrons remain in the semiconductor and increase its electrical conductivity, photoconductivity, so this effect is applied in the manufacturing of the photo-resistors.

The internal photoelectric effect with barrage layer *appears at the junction of two contaminated (doped) semiconductors or at the metal-semiconductor junction where under the action of the light is produced a photovoltaic electric field.*

This effect is considered in the manufacturing of the photo-elements (converters which transform the light or solar energy in electric energy), photodiodes and phototransistors.

The photodiode is a **pn** junction operating in reverse bias based on the internal photoelectric effect with barrage layer; the barrage layer is the depletion zone of the **pn** junction.

The photodiode is so manufactured that the **pn** junction can be lit and the reverse current depends, in a range, by the light flux intensity being proportional with it, so actually the photodiode works as a light-controlled variable resistor.

The symbol of the photodiode is presented in FIGURE 49.

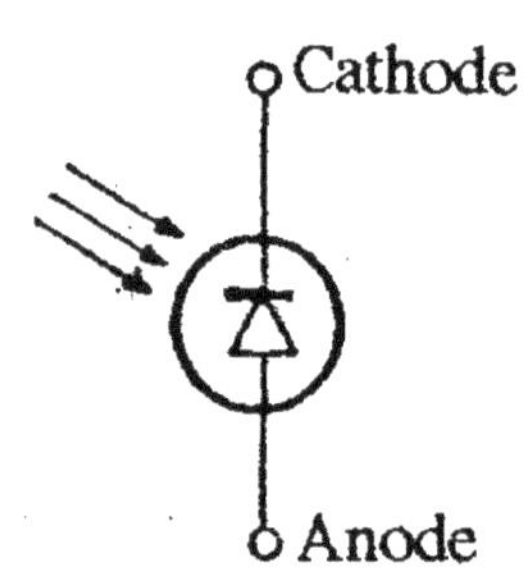

FIGURE 49 Photodiode symbol

In the circuit presented in FIGURE 50, without light through the photodiode passes the residual current called the *dark current* which for Ge manufactured diodes is 10-20 μA and for Si manufactured diodes is 1-2 μA.

When the junction is lit the current increases proportional with the illumination.

That is because the energy quanta absorbed by the semiconductor from the light flux break internal connections producing *electron & "hole"* pairs. An external electric field of a battery determines the "holes" to move to the minus of the battery and the electrons to the plus of the battery.

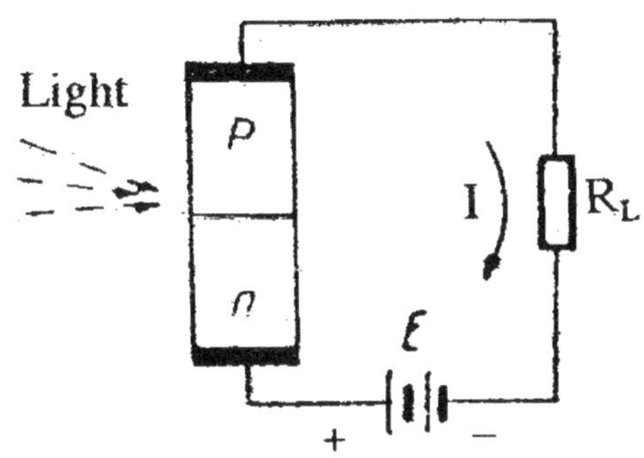

FIGURE 50 Photodiode in circuit

The photodiode is used exclusively in reverse conduction, that means in the quadrant III of its I-V characteristic curve presented in FIGURE 51.
The reverse current of the lit photodiode is called *illumination current* and has much higher values than the dark current.
The illumination current is proportional with the light flux incident on the photosensitive surface, being independent of the reverse voltage (less than the breakdown voltage).
For the photodiodes manufactured with silicon (which is preferred for photodiodes) the breakdown voltages are in the range of tens to hundreds volts.

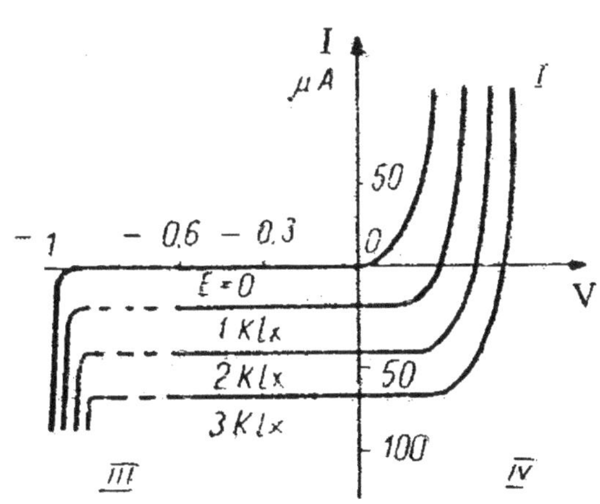

FIGURE 51 I-V characteristics of photodiode function of its illumination

If the photodiode works without external reverse bias, in the quadrant IV FIGURE 51, it can be used as photovoltaic cell (photoelectric converter) since the action of the light produces a photovoltaic electric field; a resistor between its electrodes would be passed by an electric current (very low) proportional with the external illumination E.
The intersection of the characteristic curves in the quadrant IV with the axes of the graph marks the short circuit currents I_{sc} respectively the electromotor voltages V_o (open circuit).

In comparison with the photoelectric cell, the photodiode has higher sensitivity, smaller size and greater reliability but a lower response time.
The photodiode is characterized by:

S Sensitivity - ratio of the reverse current to the illumination that produced that current (for an applied reverse voltage) (nA/lx or mA/lm)

V_R Reverse voltage

I_D Dark current - given by the manufacturer for a certain reverse voltage

t_r & t_f Rising time & falling time - intervals in which the current varies from 10% to 90% respectively from 90% to 10% of its maximum value when the photodiode is illuminated with a square pulse FIGURE 52

also:

Spectral sensitivity - the photodiode response for different light wavelengths

Directional characteristic of a photodiode - relative variation of the current with the incidence angle of the light on the sensitive surface

The photodiodes with the sensitivity of hundreds mA per lumen are used in the command/control circuits in automatic equipments (for ex. at the automatic command of the outside lighting) but the most spectacular application is in the manufacturing of the video capture tubes (image capture) which transform the information carried by the light in an electric information, those tubes being the essential part of a TV camera.

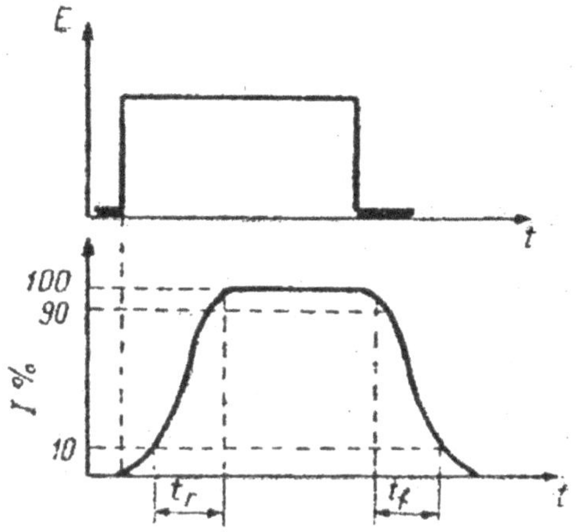

FIGURE 52 Photodiode response to the rectangular light pulse

It is interesting to present here the Schottky *photodiode*, the PIN *photodiode*, the *phototransistor*, the *photothyristor*, the *photoelectric converter* and the *solar battery*.

Schottky *photodiode*

The Schottky photodiode is the photodiode manufactured with a metallic layer on a semiconductor surface, the metal is chosen to make a rectifying contact with the semiconductor. The result is a photodiode with short switch time (< 1 ns).

PIN *photodiode*

The PIN photodiode is the photodiode which has between the **p** region P and the **n** region N an intrinsic region I. Increasing the distance between the **p** region and the **n** region the device capacitance is considerably decreased and the photodiode can work with light signals of frequency up to 1000 times higher than the usual photodiode.

Phototransistor

The phototransistor is the transistor with the collector current controlled by lighting one of the three regions (base, emitter or collector).

The base terminal is used only for the stabilisation with the temperature variation.

The main advantage of the phototransistor over the photodiode is its ten to hundred times higher sensitivity for the same photosensitive aria and the same material.
As disadvantage is the big value of the dark current and that's temperature dependence, the function of the phototransistor being affected by temperature more than the function of the photodiode.
Having in view the above mentioned things the phototransistor is used in circuits that only detect the presence or the absence of the light without any precise discrimination of the light levels.

Photothyristor

The photothyristor is the thyristor for which the gate current to trigger switching is provided by internally generated electron & "hole" pairs upon exposure to light.
The I-V characteristic curve of the photothyristor with the illumination E as parameter are given in FIGURE 53.

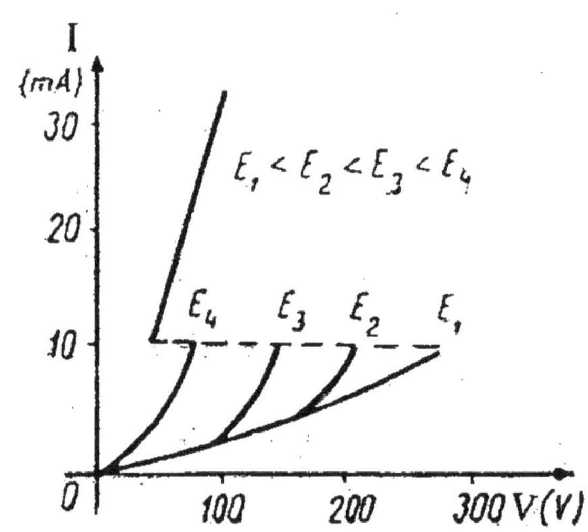

FIGURE 53 Photothyristor I-V characteristic

Photoelectric converter

The photoelectric converter is the converter that transforms directly the solar energy in the electric energy.

It is based on the optical excitation of the **pn** semiconductor junction.
The silicon photoelectric converter has the **n** region (layer) doped with pentavalent phosphor and the **p** region (layer) doped with trivalent bor.
In the barrage layer of 10^{-6}- 10^{-7} metres there are not mobile electric charge carriers.
By absorbing light quanta (photons) appear electron & "hole" pairs which under the influence of the electric field of the barrage layer are oriented, the electrons charge negatively the **n** layer, appearing a photovoltaic electric field a phenomenon called the internal photoelectric effect with barrage layer.
As materials are used: Si, CdS, CdTe, AsCa etc.
The photoelectric converter is used to estimate the illumination (lux-meter) in photography technique, and more in series/parallel connection as solar batteries.
The photoelectric converter is known too under the names:
photovoltaic converter, photoelement, photovoltaic cell, solar pile.

Solar battery

The solar battery is the battery made with photoelectric converters.
It can contain tens of thousands of elements of few square centimetres arias connected series and parallel and it is used in connection with an accumulator or storage battery so its electric energy is available also when there is not sun.
Solar energy conversion is attractive because it is inexhaustible and non-polluting.
The solar batteries are used as kW energy sources for irrigation water pumps, remote rural areas, remote weather stations, marine navigation lights, telecommunication links, satellites and cosmic ships (for ex. the Apollo missions had solar batteries of 10kW with 110 m^2 total aria of the four wings). For energy purposes, the high power solar batteries should be placed outside the filtering properties of the Earth atmosphere.
The low power solar batteries (mW) are used to supply the microelectronic circuits (watches, calculators) or the artificial organs implanted in the human body.
A maximum efficiency of 22% is obtained with photoelements made with GaAs covered with a thin AsAlGa layer.

LIGHT-EMITTING DIODE

The light-emitting diode LED is an optoelectronic device that transforms the electric energy in the electromagnetic radiation of the light.

It is a popular electronic light source based on *electroluminescence.*

The electroluminescence *is a process that changes an electrical input to a light output, the opposite of the photovoltaic effect.*

The light-emitting diode is the diode with the property of emitting light in forward conduction. It is a special **pn** junction designed for optimum electroluminescence.

The symbol of the light-emitting diode LED is presented in FIGURE 54.

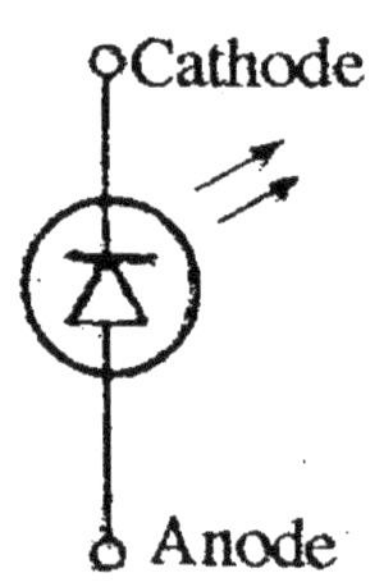

FIGURE 54 Light-emitting diode LED symbol

The LED is manufactured from compounds of gallium (Ga) (*grey metallic element discovered in 1875 belonging to the group III of the Periodic Table)* with trivalent or pentavalent elements forming GaAs, GaP, GaAsP etc.

The GaAs junctions emit radiations in the infrared range of the electromagnetic waves and introducing phosphor atoms the emitting maximum is moved to the visible range of the electromagnetic spectrum.

Nowadays are manufactured red light, yellow light or green light LEDs.

With GaP the light is red or green function of doping, with GaAsP the light is red, with SiC the light is yellow.

The electrical parameters of the light-emitting diode are identical to the general parameters of the diode, for example:

I_F maximum forward current (tens milliamperes range)

V_F opening voltage of the junction varies from 1.2V for the infrared rays emitting diodes to 3V for the green light emitting diodes

V_R maximum reverse voltage (few volts)

The parameter defining the LED quality is the *intensity of the emitted electromagnetic radiation that for visible is noted* I_v *measured by millicandela (mcd) and for infrared is noted* I_e *measured* by *mW/steradian.*

Either I_v or I_e vary linear with the device current I for a large range of that's values.

The I_v- I characteristic curve is presented in FIGURE 55.

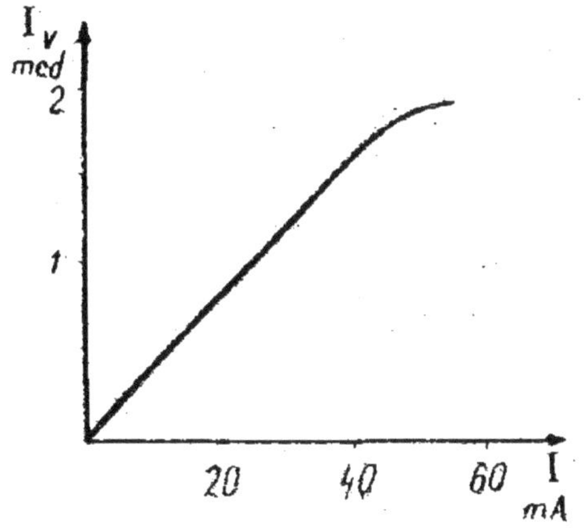

FIGURE 55 Emitted light intensity I_v function of forward electric current I_F for LED

However for high currents the dependence is not more linear, because the device's temperature increases and also because of the quantum effects which appear at high densities of the junction current.

The light-emitting diode is also characterised by its *spectral characteristic curve* (intensity of emitted light / frequency of emitted light) and its *conversion efficiency.*

Alike to the devices which work absorbing light, for LEDs are defined two response times: rising time t_r and falling time t_f.

They represent the time intervals for the radiation or light intensity to rise from 10% to 90% respectively to fall from 90% to 10% of its maximum value when to the LED is applied a rectangular current pulse (ns for infrared emission).
On the I-V characteristic curve of the LED, the working point M is chosen in the first quadrant because the light is emitted when the device is forward biased FIGURE 56.

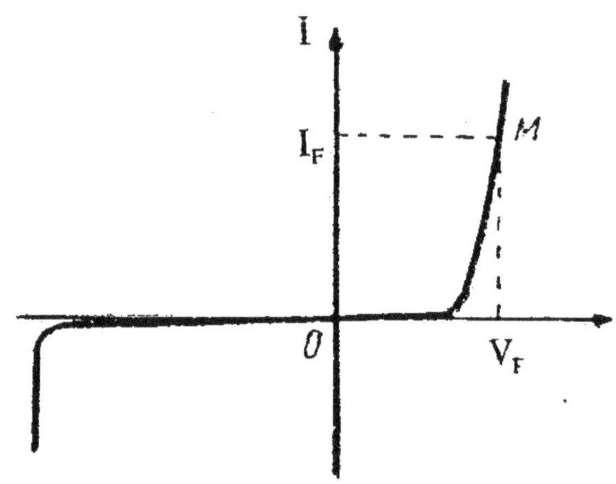

FIGURE 56 I-V characteristic of LED

The current I_F corresponding to this point is limited by a resistor R in series with the LED as in FIGURE 57. The resistor value is calculated function of the current I_F and the voltages V_F and V:

$R = (V - V_F)/I_F$ where V_F = forward voltage on LED V = supplier voltage

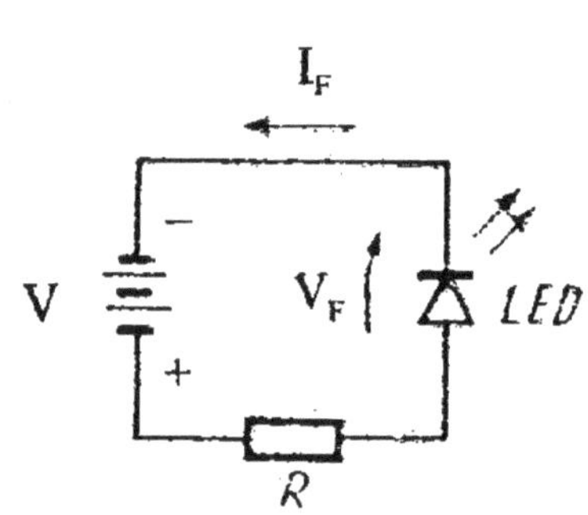

FIGURE 57 LED in circuit

The low current needed “to light up“ the LED made it to be preferred to the incandescent bulb, and the low operating voltage made to be preferred to the bulb with electric discharge in inert gas.
A special technology could manufacture a light emitting diode with laser-like light features: advanced monochromatic, directness, coherence and high intensity.
That device is the *laser diode* or the *injection laser*.

The LED is a practical light source in the modern electronics.
The light-emitting diodes are used for example to built the optical couplers or opto-couplers.
An opto-coupler consists of a light source that converts an electrical signal to a light signal and a photodetector that converts the light signal back to the electrical signal, the two parts laying face to face at small distance in the same capsule; it presents the advantage of complete electrical isolation avoiding the problem of impedance matching and the case of circuit noise.
There is a large diversity of opto-couplers. Theoretically it is possible to use any photo-emitting device (electric bulb, LED) and any photo-detector device (photovoltaic cell, photo-resistor, photodiode, phototransistor, photo-thyristor) with the condition that the emission spectrum to correspond to the spectral sensitivity of the light detector.

In FIGURE 58 are presented six possible structures for optical couplers:

a. incandescent bulb - photoresistor
b. LED - photodiode
c. LED - phototransistor
d. LED - photovoltaic cell
e. LED - photothyristor
f. LED - photodarlington (phototransistor in Darlington connection)

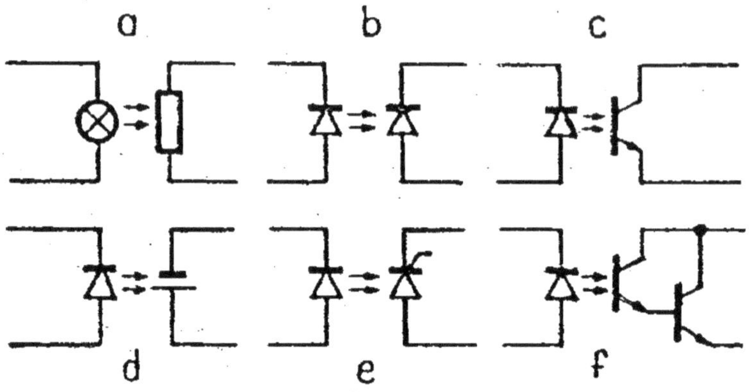

FIGURE 58 Opto-couplers

LEDs are used on a large scale as "power on" indicators and as displays for calculators, digital voltmeters, frequency counters, etc being placed more together in seven-segments displays as in FIGURE 59.

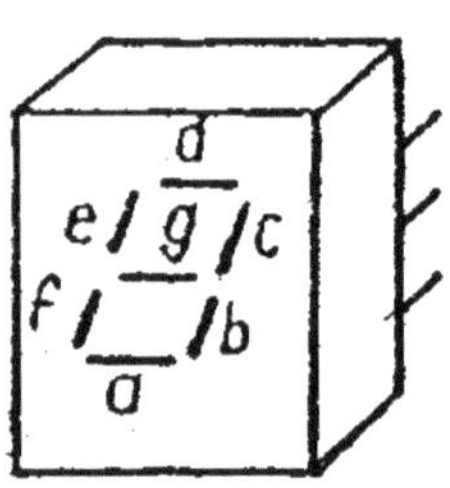

FIGURE 59 Seven-segments LED display

The seven-segments display uses seven LED segments labelled *a, b, c, d, e, f, g* which can be lit in different combinations to form any number from "0" through "9".
Each segment is a LED using for example 1.6V at 20mA.
To control the seven-segment LED display is used the electronic circuit presented in FIGURE 60. LEDs with common anode are supplied each through a resistor limiting the current; with the help of the silicon diodes, LEDs are switched selectively function of the position of the switch K.

Each position of K is translated (decoded) in a combination of lit segments that represent a figure; the circuit from FIGURE 60 is called *decoder*.
In circuit could be used the digital device ROLL77 and the diodes 1 N4001 or similar.

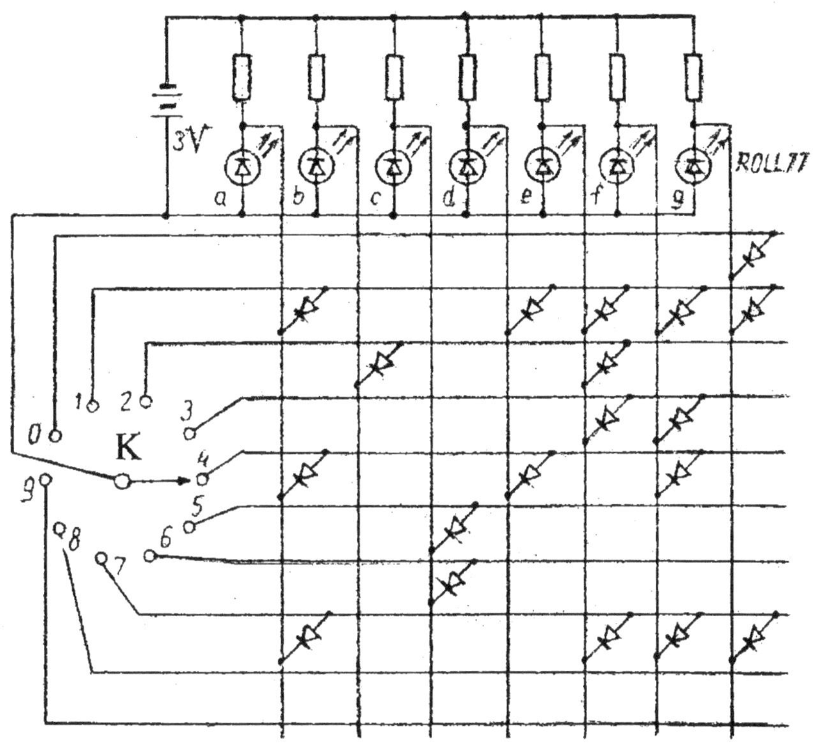

FIGURE 60 Decoder - command circuit for the seven-segment LED display

PART II
TRANSISTORS

THE TRANSISTOR

The intensive study of the semiconductors properties begins in the year 1948 after the American physicists John Bardeen and Walter Brattain announced the discovery of the point-contact transistor, the first solid amplifier.

The discovery occurred when they actually were searching for a field effect transistor.

In the next year the American physicist and inventor William Shockley realised the bipolar transistor with junctions, presenting the theory of this new electronic device, which revolutionized the science and the engineering.

The word "transistor" comes from ***transfer resistor*** being a name coined at the beginning of the semiconductor era for the bipolar transistor instead of keeping the old equivalent to the vacuum tube "triode" as in the case of "diode".

Like the vacuum tube (thermionic valve), the transistor is used to produce, amplify, commute and rectify the electric waves and impulses.

Roughly, the transistor is as size a thousandth of the volume, has a hundredth of the weight and needs a tenth of the power of the vacuum tube.

Typical semiconductors as germanium and silicon are basic elements for the manufacturing of the transistors.

Depending on the passing electric current, the transistors are divided in:

1. *bipolar transistors* where the passing electric current is the moving of both, electrons and "holes" - typical example being:
   Bipolar Transistor with junctions or *Bipolar Junction Transistor* BJT
2. *unipolar (monopolar) transistors* where the passing electric current is the moving of either the electrons or the "holes" - typical example being:
   Field Effect Transistor FET with the two important representatives:
   - *FET with junction gate* J-FET
   - *FET with insulated gate* MIS-FET (Metal-Insulator-Semiconductor FET) or MOS-FET (if the insulator is SiO_2)

BIPOLAR TRANSISTOR

Representative for the bipolar transistor is the *Bipolar Junction Transistor* BJT or the *Bipolar Transistor with junctions.*

BJT represents the most used command active device usually called the *transistor*.

In essence is a **pnp** or **npn** structure, a semiconductor monocrystal divided in three distinct regions by impurities of type **p** or **n** introduced in them, regions separated by two **pn** junctions. Depending on their role in the transistor function the three regions of the mono-crystal are called *emitter, base* and *collector* and the two junctions are called *emitter junction* and *collector junction, see* FIGURE 1.

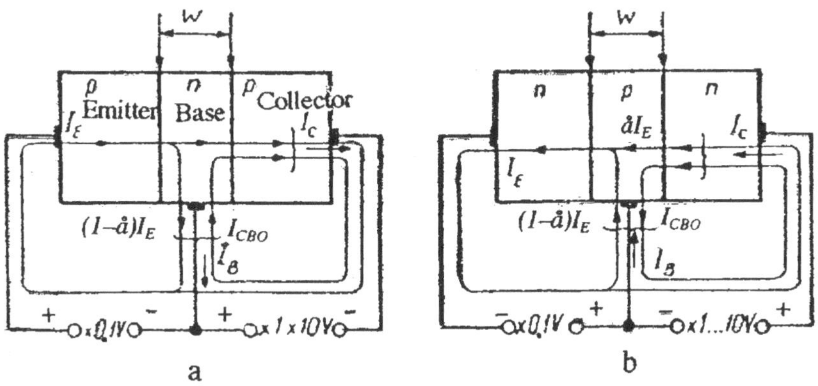

FIGURE 1 a. pnp Transistor b. npn Transistor

Illustratively the transistor can be seen as a "sandwich" with one **n** type layer between two **p** type layers (**pnp** transistor) or with one **p** type layer between two **n** type layers (**npn** transistor).

In FIGURE 2 is presented a section through the transistor device where:

a. emitter contact

b. base contacts

c. collector contact on the inferior side

d. collector region

e. spatial regions

f. base region

g. emitter region

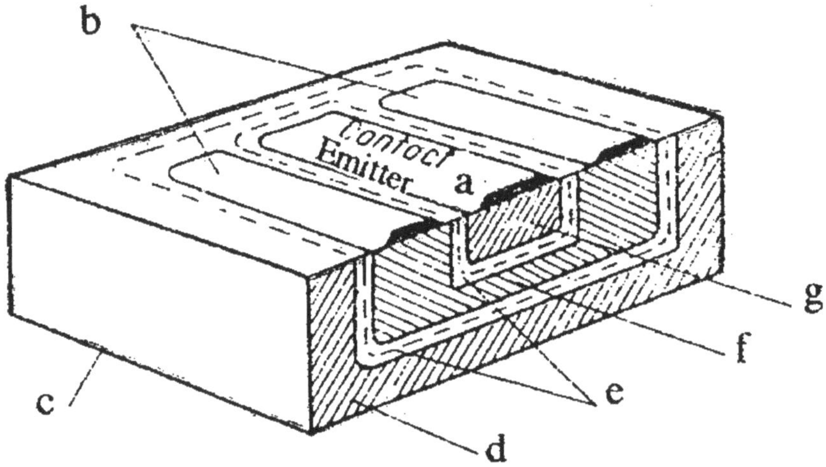

FIGURE 2 Transistor structure

The standard symbols for **pnp** and **npn** bipolar transistors are presented in FIGURE 3. The emitter terminal is represented with an arrow oriented in the conventional direction of the emitter current flow when the transistor works in normal active regime, with the emitter junction in forward conduction and the collector junction in reverse conduction.

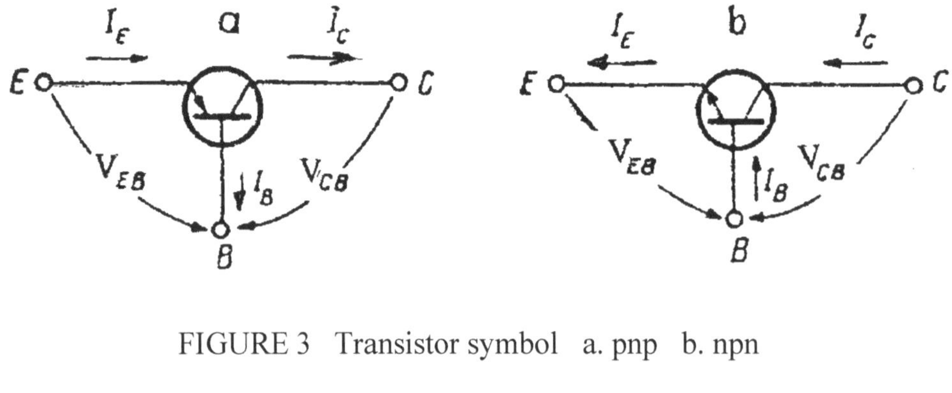

FIGURE 3 Transistor symbol a. pnp b. npn

To analyse the physical function of the transistor generally it is used the **pnp** transistor. The fundamental behaviour of the **npn** transistor is similar, the voltages and the currents having opposite signs.
The notations for the currents and voltages at the transistor terminals are standard notations. For static regimes the standard notations use upper case letters using for subscript also upper case letters. So I_E and V_{EB} are constant or change very slow so the relations between them are in essence the same with the relations for the static regimes.
For instantaneous/dynamic regimes (signals) the standard notations use lower case letters using for subscript also lower case letters but for a total value the subscript has upper case letters.

In normal function the emitter junction is forward biased and is in forward conduction and the collector junction is reverse biased and is in reverse conduction; it is said that the transistor functions in the normal active regime NAR.
If the emitter junction is reverse biased and is in reverse conduction and the collector junction is forward biased and is in forward conduction, it is said that the transistor functions in the reverse active regime RAR.
In the initial transistor theory is considered that the phenomena in transistor depend on a single coordinate along a perpendicular axis on junctions; the assumption is closer to the reality if the junctions are plan and surface effects reduced.
The base region is very thin. In NAR if the base thickness *w* would be big, the two junctions would be independent each other and the emitter junction would bc passed by the forward current and the collector junction would be passed by the reverse current.
Because the base is manufactured very thin [w = tenths of mm << diffusion length of the minority electricity carriers] and very little doped, the emitter junction current will pass also the collector junction, phenomenon called the transistor effect.
The proportion of the emitter current (which is the same with the emitter junction current) that becomes collector current is defined by the *emitter-collector current gain* å:

å = collector current determined by emitter current / emitter current

The part of the emitter current that becomes base current will define the *base-collector current gain* ß as the ratio of the emitter current that becomes collector current to the emitter current that becomes base current:

ß = collector current determined by emitter current / base current determined by emitter current

Between å and ß there are the relations $å = ß / (1+ ß)$ $ß = å / (1- å)$
$å = 0.9 - 0.999$ $ß = 10 - 10^3$

The residual current of the collector junction i_{CBO} contributes to the total collector current and to the total base current.
The fundamental relation of the transistor function is:

$$i_C = å\ i_E + i_{CBO}$$

Notations and relations:

i_{CBO} residual current of base-collector junction (i_C with emitter in open-circuit $i_E = 0$)
[~ 10 μA for Ge transistors / ~ 0.1 μA for Si transistors]

i_E	emitter current	$i_E = i_C + i_B$
i_B	base current	$i_B = (1–å)i_E – i_{CBO}$
i_C	collector current	$i_C = å\ i_E + i_{CBO}$
		$i_C = ß\ (i_B + i_{CBO}) + i_{CBO} = ß\ i_B + (ß+1)\ i_{CBO}$

The considering that the collector current is determined by the voltage on the emitter junction would define the *transconductance.*
Following the transistor effect phenomenon the current that travels between emitter and collector (perpendicular on junctions) is controlled by the voltage on the emitter junction corresponding to the forward conduction of the emitter junction and is independent of the voltage on the collector junction so long the collector junction is in reverse conduction. The collector current is little less than the emitter current and the current that travels between emitter and base (which is also independent of the voltage on the collector junction so long the collector junction is in reverse conduction) is much less than both.

Theoretically at the collector terminal, the transistor can be seen as a controlled *constant current generator*.
The control can be done in more ways:

a. The collector current is controlled by the emitter-base voltage and increases rapid by increasing the value of the forward voltage on the emitter junction.
In conformity with the study of the **pn** junction it is expected that this dependence to be exponential $I_C \approx k^{VEB}$ that is confirmed by FIGURE 4 where in semi-logarithmic co-ordinates is presented the dependence of the collector current on the emitter-base voltage for a Ge transistor. Because of the exponential relationship the bipolar junction transistor has a higher transconductance than the field effect transistor and is seen as a *transconductance device.*

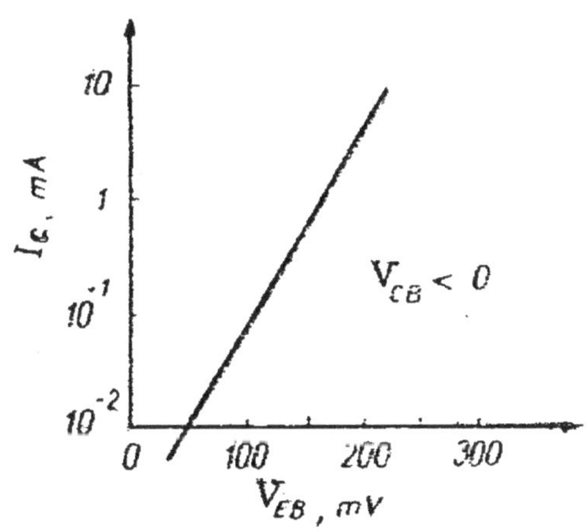

FIGURE 4 Collector current I_C controlled by emitter-base voltage V_{EB}

b. The collector current is controlled by the base current.
This viewpoint generated the graphical representation presenting the collector current independent of the collector voltage with the base current as parameter, see FIGURE 5.

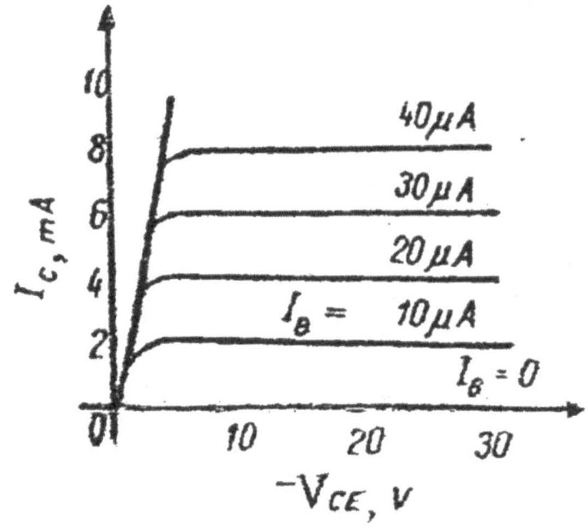

FIGURE 5 Collector current I_C controlled by base current I_B

c. The collector current is controlled by the emitter current.

In general this viewpoint is not useful for the electric description of the device because for the majority circuits with transistors the "input signal" is applied to base and the emitter and the collector currents are approximately equal.

Because the relation between the emitter current and the emitter-base voltage has the same representation as the I-V characteristic curve of the ideal diode with **pn** junction, the behaviour of the transistor at the emitter-base terminals can be described using an ideal diode with **pn** junction and at the collector-base terminals the exponential dependence of the collector current on the emitter-base voltage can be described with a dependent or controlled constant current generator, see FIGURE 6.

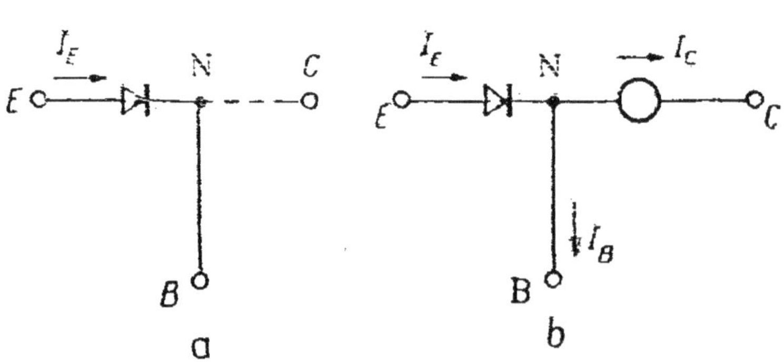

FIGURE 6 Equivalent circuit of transistor in static active regime a. partial b. complete

Considering that the control is done by the emitter-base voltage V_{EB} in the N point of FIGURE 6 can be written the relations:

$I_E - I_B - I_C = 0$

$I_E = I_B + I_C$

$I_E = I_{EBO} \{\exp [eV_{EB} /(kT)] - 1\} = I_{EBO} [\exp (V_{EB} /V_T) - 1]$

$I_B = \partial I_E$ (the base current I_B is expressed as a fraction of the emitter current I_E)

$I_B = \partial I_{EBO} [\exp (V_{EB} /V_T) - 1]$

$I_C = (1-\partial) I_E = (1-\partial) I_{EBO} [\exp (V_{EB} /V_T) - 1]$

I_{EBO} residual current of the emitter-base junction (with collector in open circuit)

$V_T = kT/e$ thermal voltage (25 mV for Ge at 17 °C)

e electron electrical charge $1.602177\ 10^{-19}$ C

k Boltzman constant $1.380658\ 10^{-23}$ J/K

T temperature given in Kelvin degrees K

∂ coefficient called *base defect* and is approximately $10^{-1} \ldots 10^{-3}$. Its value is difficult to control in design and manufacturing of the transistors. A group of the same type transistors can have ∂ even ten times different.

The ratios of the currents' modules, which are the current static gains å and ß are independent of V_{EB} and I_{EBO} depending only on ∂:

$$| I_C | / | I_E | = 1 - \partial = å$$

$$| I_C | / | I_B | = (1 - \partial) / \partial = ß$$

The above presentation is based on two assumptions:

1. The emitter junction is in forward conduction with a voltage drop on it much higher than $V_T = kT/e$
2. The collector junction is in reverse conduction with a voltage drop on it in absolute value much higher than $V_T = kT/e$

These conditions define the limits of the active regime for which are valid the equivalent circuits of FIGURE 6.

A way to leave the active regime is to bias reverse the emitter junction. In that case the transistor is in the cut-off (OFF) state where the currents are very low and independent of the voltages applied on junctions.

The behaviour of the transistor in the OFF state can be simulated with an open circuit as in FIGURE 7a.

The transistor can leave the active regime also if the collector junction is forward biased and in that case the device is in the saturation regime. Both junctions are forward biased, the voltage drop on them is small and relatively independent of the currents.

The electric behaviour of a transistor in saturation can be approximated with a short circuit as in FIGURE 7b.

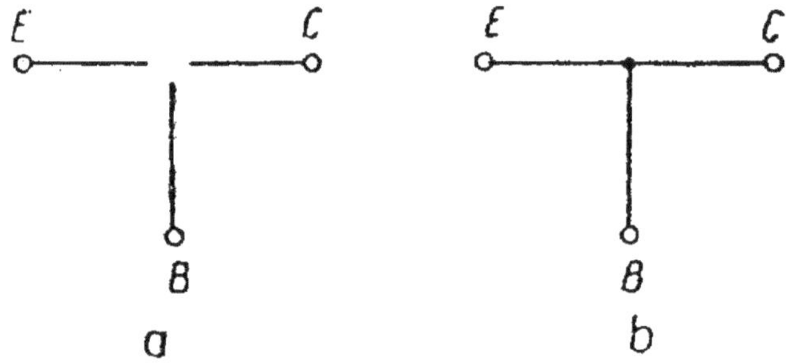

FIGURE 7 Equivalent circuits for a. cut-off regime b. saturation regime

If the input signal is applied to the emitter-base terminals and the output signal is at the collector-base terminals the transistor works in the common-base configuration CB.

If the common terminal is the emitter the circuit operates in the common-emitter configuration CE.

If the common terminal is the collector the circuit operates in the common-collector configuration CC.

For the common base configuration CB the input parameters are the emitter current I_E and the emitter-base voltage V_{EB} and the output parameters are the collector current I_C and the collector-base voltage V_{CB} FIGURE 8a.

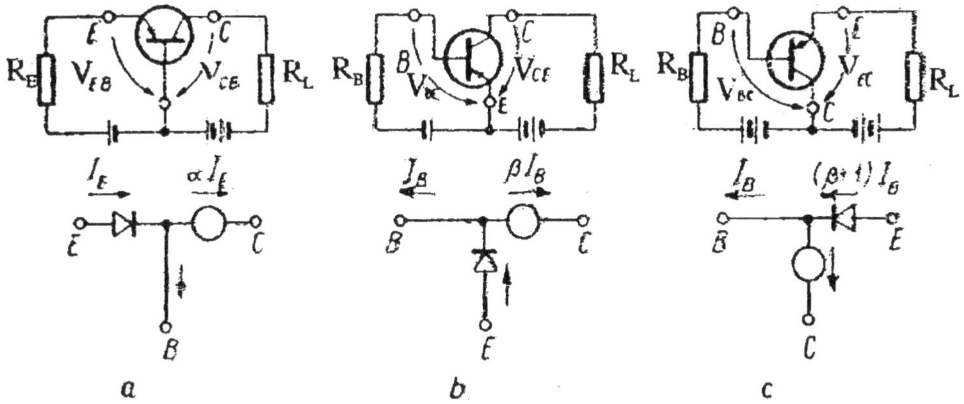

FIGURE 8 Transistor configuration circuits
a. common-base b. common-emitter c. common-collector

For the common emitter configuration CE the input parameters are the base current I_B and the base-emitter voltage V_{BE} and the output parameters are the collector current I_C and the collector-emitter voltage V_{CE} FIGURE 8b.

For the common collector configuration CC the input parameters are the base current I_B and the base-collector voltage V_{BC} and the output parameters are the emitter current I_E and the emitter-collector voltage V_{EC} FIGURE 8c.

The choice of the configuration depends on the value of the input and output impedances or resistances and of the voltage, current and power gains.

The most used is the common emitter configuration CE.

FIGURE 9 illustrates how the **pnp** bipolar transistor is used in a circuit as control and amplification device.

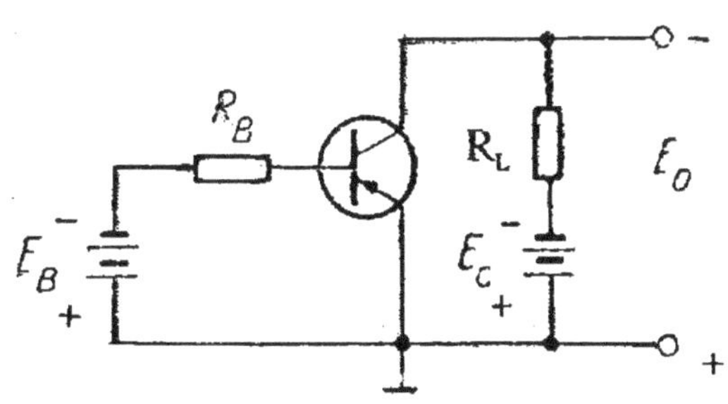

FIGURE 9 Amplifier with transistor in common-emitter configuration

The transistor collector and emitter belong to an output circuit together with a direct voltage source E_C and the load resistor R_L.

The input circuit connected between the base and the emitter is composed by the direct voltage source E_B and the resistor R_B.

Since the emitter belongs to both the input circuit and the output circuit, the transistor is in a typical common-emitter configuration.

The circuit can be represented with the static model for the normal active regime from FIGURE 7b where the opened diode was replaced by a short circuit.

The result is the electric circuit presented in the FIGURE 10.

$I_B = E_B / R_B$

$I_C = ß \cdot I_B$

$E_o = E_C - I_C R_L = E_C - ß I_B R_L = E_C - ß E_B R_L / R_B$

The previous relation shows that the output voltage of the amplifier E_O is more sensitive to the variation of the E_B voltage than E_C voltage.

Indeed: $\Delta E_o / \Delta E_C = 1$ for E_B constant *and* $\Delta E_o / \Delta E_B = -ß R_L / R_B$ for E_C constant

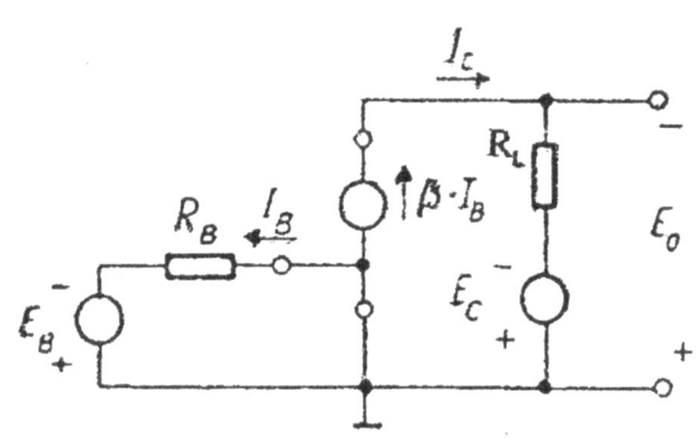

FIGURE 10 The circuit of FIG 9 with the transistor replaced by its equivalent circuit

Considering the usual values $R_L = R_B = 1k\Omega$ and $ß = 100$ E_o will vary with 100 mV when E_B will vary 1mV and only with 1mV when E_C will vary 1mV.

That means that the above circuit has the useful property to amplify the voltage with the *voltage amplification* A_V.

$$A_V = \Delta E_o / \Delta E_B = - ß R_L / R_B$$

Note: amplification = gain

The same circuit has also the property to amplify the current. The variation of the output current produced by the variation of the input current is the *current amplification* A_I .

$$A_I = \Delta I_C / \Delta I_B = \beta$$

For the values introduced above a variation of 1μA of the input current ΔI_B will produce 100μA variation of the output current ΔI_C.

The transistors are not used in the electric circuits because they can amplify the electric voltage performing the voltage amplification A_V or because they can amplify the electric current performing the current amplification A_I , each could be realised separately with a transformer, but because *they can amplify the voltage and the current simultaneously in the same circuit meaning that they can amplify the electric power performing the power amplification* A_P .

The power supplied by the input source E_B : $P_{input} = E_B I_B = I_B^2 R_B$

The power supplied by the output source E_C : $P_{output} = E_C I_C = E_o I_C + I_C^2 R_L = P_D + P_L$

The first term from the right side of the last equation is the power given to the transistor (and dissipated by that as heat) and the second term from the right side is the power given to the load resistor. The ratio of the power developed in the load and the power supplied by the input source is the *power amplification* A_P .

$$A_P = P_L / P_{input} = (I_C / I_B)^2 \cdot R_L / R_B = \beta^2 R_L / R_B$$

For the values given above the power amplification A_P is 10^4.

It is observable that the power amplification A_p is the absolute value of the multiplication of the voltage amplification A_V and the current amplification A_I respectively:

$$A_P = | A_V \cdot A_I |$$

The previous demonstration shows that the transistor is a suitable device to control the power. The power delivered to the load resistor by the supplier E_C is controlled by the input variable E_B. More, the control is realised using a low input power.

EBERS-MOLL EQUATIONS

It is interesting to present here the *Ebers-Moll equations* which permit the calculation of the transistor currents using its parameters.

The equations obtained by J.J. Ebers and J. L. Moll describe the behaviour of the transistor for any bias of the junctions , for all three working regimes (active, cut-off, saturation).

They also lead to an equivalent circuit in stationary regime of the transistor, valid for any bias applied to the two junctions (with current generators which underline the transistor effect).

$$\text{å}_N \cdot I_{EBO} = \text{å}_R \cdot I_{CBO}$$

$$I_C = -\text{å}_N I_{EBO}[\exp(V_{EB}/V_T) - 1]/(1 - \text{å}_N \text{å}_R) + I_{CBO}[\exp(V_{CB}/V_T) - 1]/(1 - \text{å}_N \text{å}_R)$$

$$I_E = I_{EBO}[\exp(V_{EB}/V_T) - 1]/(1 - \text{å}_N \text{å}_R) - \text{å}_R I_{CBO}[\exp(V_{CB}/V_T) - 1]/(1 - \text{å}_N \text{å}_R)$$

V_T is the thermal voltage.

$V_T = kT/e = 25$ mV for Ge transistor at the temperature 17°C

$V_T = nkT/e = 25...52$ mV for Si transistor at the temperature 17°C

Note: for the values V_j of V_{EB} and V_{CB} higher than 0.1 V $\exp(V_j/V_T) >> 1$ *and* $\exp(-V_j/V_T) << 1$

I_{CBO} = residual collector current

I_{EBO} = residual emitter current

$\text{å}_N = \text{å}$ for normal active regime of transistor

$\text{å}_R = \text{å}$ for reverse active regime of the transistor

The Ebers-Moll equations characterise the intrinsic transistor that means that V_{EB} and V_{CB} are voltages applied to the junctions not to the terminals.

They are used with good precision either for the Ge transistors or the Si transistors in calculation relating to biasing circuits, direct current amplifiers, choppers and commutation circuits, after experimentally determination of the four parameters: I_{CBO} I_{EBO} å_N å_R

STATIC REGIME

In an electronic circuit a transistor should work at a certain collector current I_C and a certain collector-emitter voltage U_{CE} in consequence the base current I_B has a certain value and so on; it is said that the transistor works in the static or stationary regime in a certain *operating point.*

Between the working variables of the transistor in the static regime there are complex inter-depending relations that in a graphical representation become the *static characteristic curves.*

The usual static characteristic curves are:
- input current / input voltage
- input current / output voltage
- output current / input voltage
- output current / output voltage
- output current / input current
- output voltage / input voltage

In catalogues are given families of experimental static characteristic curves with a third variable as parameter.

The function of an electric amplifier with transistor is determined if there are given four of its variables: input current I_1, input voltage V_1, output current I_2 and output voltage V_2.

A single family of static characteristic curves gives the inter-dependence between three of the variables.

To determine the fourth variable it is necessary the second family of curves.

For a transistor there are more families of static characteristic curves, from which only two are independent, the others can be deduced from that.

In FIGURE 11 are represented for the transistor in CE configuration the input static characteristic curves $I_B = f(V_{BE}, V_{CE})$ and the output static characteristic curves $I_C = f(V_{CE}, I_B)$ which form a pair of two independent families of characteristic curves.

In this case $I_1=I_B$ $V_1=V_{BE}$ $I_2=I_C$ $V_2=V_{CE}$.

From these experimental graphics is observed that the base current and the collector current vary with the collector voltage; that is because of the real values of the internal electrical resistances of the transistor.

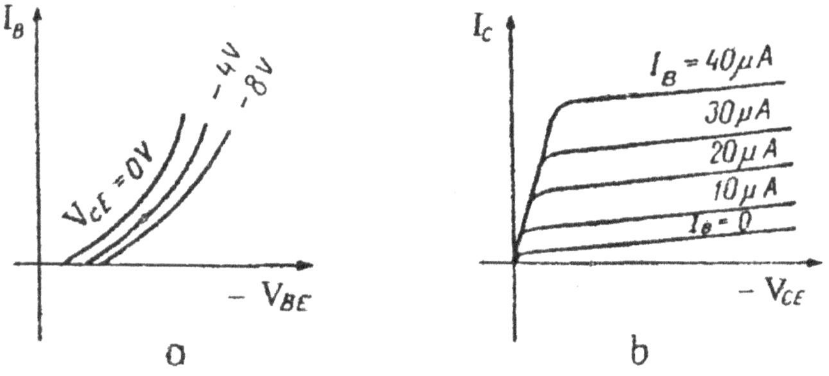

FIGURE 11 I-V static characteristics for EC configuration of transistor
a. input characteristics b. output characteristics

Practically the transistor does not function like a perfect constant current generator.
A certain operating point can be chosen on the static characteristic curves and is obtained in practice with the help of the bias circuit.
If the operating point is not specified, it is recommended to choose that for which are given the small signal parameters in catalogues.
In FIGURE 12 are given two examples of bias circuits with serial resistor and with voltage divider in base.
The circuit from FIGURE 12a is known as bias circuit with constant base current, and being given V_{CE} , I_C , E_C , I_B , choosing R_E (from sensitivity conditions) and approximating $I_E \approx I_C$ and $V_{BE} << E_C$ will result the values of R_B and R_C :

$$R_B = (E_C - I_C R_E)/I_B \qquad R_C = (E_C - V_{CE})/I_C - R_E$$

For the circuit from FIGURE 12b being given V_{CE}, I_C, E_C, V_{BE} , choosing R_E and approximating $I_E \approx I_C$ and $I_E = 0.1...0.2\ I_D$ (I_D = divider current) will result the values of R_1 , R_2 and R_C:

$$R_1 = (E_C - V_{BE} - R_E I_C)/I_D \qquad R_2 = (V_{BE} + R_E I_C)/I_D \qquad R_C = (E_C - V_{CE})/I_C - R_E$$

It is important to give attention to the bias resistor of the base that can be short-circuited by the relative low input static resistance of the transistor.

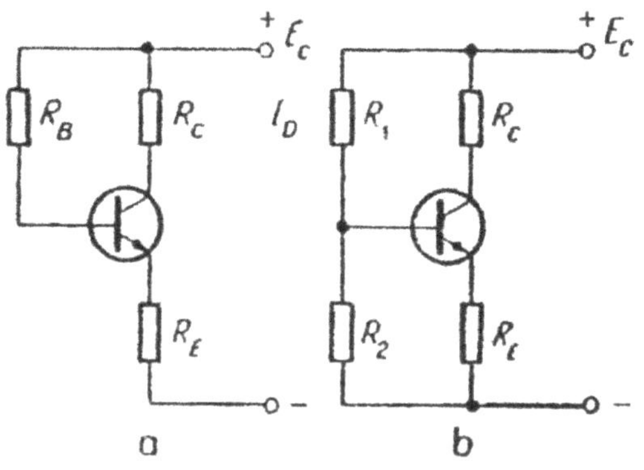

FIGURE 12 Bias circuits with serial resistor in base (a) and voltage divider in base (b)

In the design of the bias circuits, are taken into consideration the following variations with the temperature (leading to the unwanted change of the operating point):

1. Variation of the residual collector current I_{CBO} with the temperature

For the majority of the transistors it is accepted in calculation a variation law of I_{CBO} as:

$$I_{CBO}[(25 \pm \Delta T)^\circ C] = I_{CBO}(25^\circ C)\, e^{\pm a\Delta T}$$

$1/a = 12.5\ ^\circ C$ for Ge transistor

$1/a = 8\ ^\circ C$ for Si transistor

In general is counted that I_{CBO} doubles at each 9°C for Ge transistor and at each 6°C for Si transistor.

In the design of the biasing circuits the minimum value of the I_{CBO} is considered zero, and its maximum value is taken from the catalogue characteristic $I_{CBO} = f(T_j)$.

From FIGURE 13 is observed that the I_{CBO} effect is negligible for the Si made transistor if the junction temperature is under 100°C and the emitter current higher than 1μA.

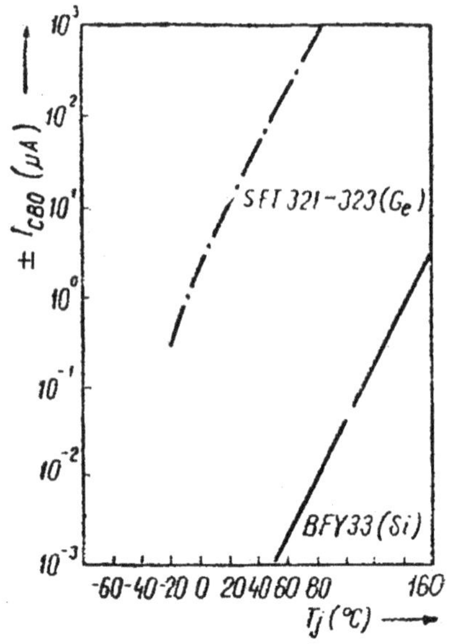

FIGURE 13 Variation of I_{CBO} with temperature for npn Si transistor (BFY 33) and for pnp Ge transistor (SFT 321-323)

2. Variation of the base-emitter voltage V_{BE} with the temperature

In the normal active regime the base-emitter voltage is around 0.2V for Ge transistor and 0.6V for Si transistor and decreases approximately 2.2mV/°C when maintains constant the emitter current FIGURE 14.

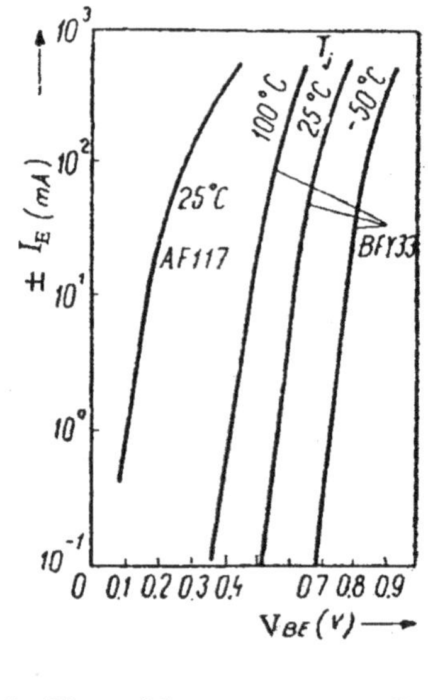

FIGURE 14 Variation of I_E-V_{BE} with temperature for npn Si transistor (BFY 33)

3. Variation of the static current gain ß with the temperature

In catalogues are specified the maximum value and the minimum value at 25°C. These values are affected by the temperature how can be seen in the $ß/T_j$ normalised characteristic curves FIGURE 15.

In the absence of the characteristic curves it is possible to calculate the gain taking in consideration the relation:

$$ß[(25 \pm \Delta T)°C] = ß(25°C)\,(1 \pm \Delta T/ç)$$

ç = 100°C for the Ge transistor

ç = 50°C for the Si transistor

For the low power transistors manufactured with silicon, the constants are given informatively because in the bias circuits design the variation of the I_{CBO} and ß with the temperature is neglected.

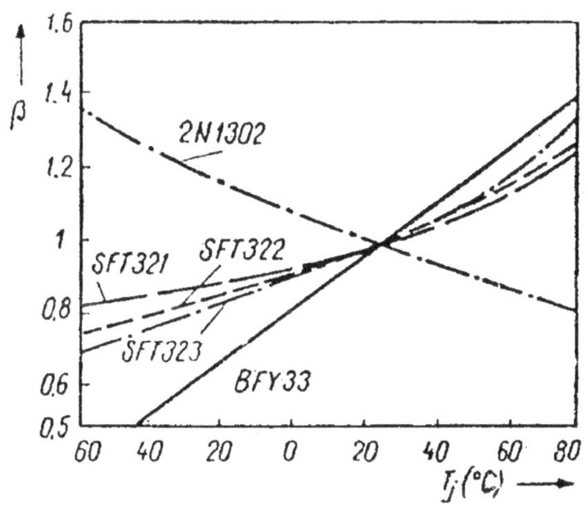

FIGURE 15 Variation of current gain ß with temperature

To stabilize the static operating point there are used two types of thermal stabilizing methods:

Linear methods

Non-linear methods (compensations)

Linear methods use the linear circuit elements and the negative reaction.

Non-linear methods use the non-linear circuit elements whose variation with temperature compensate for that of the operating point.

An example of linear method is the use of the resistor R_E (linear circuit element) which introduces a negative reaction in direct current, presented in FIGURE12.

More the temperature increases, more I_E increases, more emitter potential decreases, more V_{BE} decreases, more I_E decreases.

Higher the electrical resistance value of R_E better the thermal stabilization.

Examples of non-linear methods are the use of the diode and the thermistor (non-linear circuit elements) as in FIGURE 16.

The diode biased with 0.5-1V in reverse conduction has a reverse current equal to I_{CBO} whose variation with the temperature compensates for the variation of I_{CBO} with the temperature, see FIGURE 16a.

The thermistor whose electric resistance decreases when the temperature increase, decreases the negative potential of base when the temperature increases, maintaining I_E approximately constant, see FIGURE 16b.

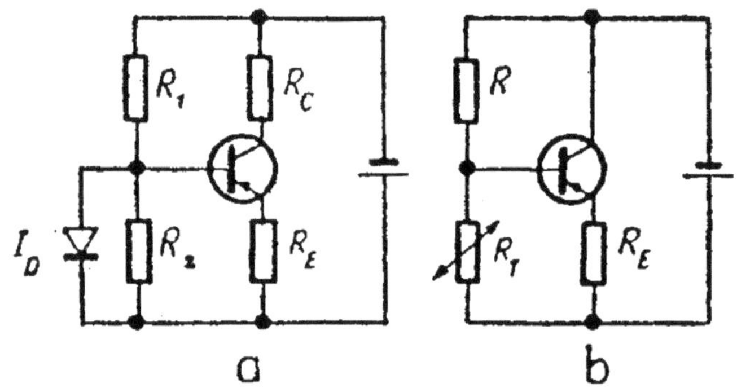

FIGURE 16 Bias circuits with thermal compensation
a. with diode b. with thermistor

DYNAMIC REGIME

In the previous chapter was taken in consideration the transistor function at a certain bias in the stationary regime.
If to the transistor are applied biases varying in time so that in every moment it is in a stationary state determined by the bias value at that moment (semi-stationary state) that corresponds to the transistor function in the variable or dynamic regime at low frequencies.
In the analysis of the amplification of low frequency electrical signals using circuits with transistors are distinguished two types of problems:

High amplitude signal amplification (final power circuits) where the transistor has non-linear behaviour

Low amplitude signal amplification where the transistor behaves as a linear electrical quadripole

The problem of big signals of low frequency is solved with the static characteristics.

The transistor, in general a non-linear device, can be considered linear for small amplitude signals and the most suitable method to analyse the transistor in linear regime proofed to be the electrical quadripole method.
Considering the transistor a quadripol as in FIGURE 17, it establishes relations between the currents and voltages at the input and the output of the transistor, i_1 v_1 and i_2 v_2, using four parameters given in *Table 1*.

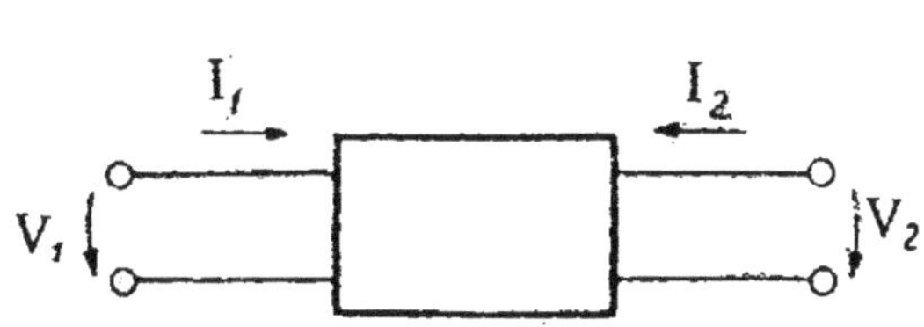

FIGURE 17 Active linear quadripole

With the four parameters it is possible to calculate the main properties of the amplifiers for small amplitude signals.

The four parameters result by writing in all possible ways the relations between v_1 i_1 v_2 i_2 as in *Table 1*.

There are six possibilities to write these relations from which three are applicable in measurements:

Considering i_2 v_2 independent variables
Considering i_1 v_1 independent variables
Considering v_1 i_2 independent variables
Considering i_1 v_2 independent variables will result the "h" (hybrid) parameters (representing impedances and admittances)
Considering v_1 v_2 independent variables will result the "y" parameters (representing admittances)
Considering i_1 i_2 independent variables will result the "z" parameters (representing impedances)

Table 1

Relations between the four variables of the equivalent quadripole

Independent Variables	General Equations	Quadripol Equations	Resulted Parameters
i_1, i_2	$v_1=f_1(i_1,i_2)$ $v_2=f_2(i_1,i_2)$	$v_1=z_{11}i_1+z_{12}i_2$ $v_2=z_{21}i_1+z_{22}i_2$	$z_{11}=v_1/i_1\ _{i2=0}$ $z_{12}=v_1/i_2\ _{i1=0}$ $z_{21}=v_2/i_1\ _{i2=0}$ $z_{22}=v_2/i_2\ _{i1=0}$
v_1, v_2	$i_1=f_3(v_1,v_2)$ $i_2=f_4(v_1,v_2)$	$i_1=y_{11}v_1+y_{12}v_2$ $i_2=y_{21}v_1+y_{22}v_2$	$y_{11}=i_1/v_1\ _{v2=0}$ $y_{12}=i_1/v_2\ _{v1=0}$ $y_{21}=i_2/v_1\ _{v2=0}$ $y_{22}=i_2/v_2\ _{v1=0}$
i_1, v_2	$v_1=f_5(i_1,v_2)$ $i_2=f_6(i_1,v_2)$	$v_1=h_{11}i_1+h_{12}v_2$ $i_2=h_{21}i_1+h_{22}v_2$	$h_{11}=v_1/i_1\ _{v2=0}$ $h_{12}=v_1/v_2\ _{i1=0}$ $h_{21}=i_2/i_1\ _{v2=0}$ $h_{22}=i_2/v_2\ _{i1=0}$
v_1, i_2	$i_1=f_7(v_1,i_2)$ $v_2=f_8(v_1,i_2)$		
i_1, v_1	$i_2=f_9(i_1,v_1)$ $v_2=f_{10}(i_1,v_1)$		
i_2, v_2	$i_1=f_{11}(i_2,v_2)$ $v_1=f_{12}(i_2,v_2)$		

The equivalent electrical circuits corresponding to the parameters “h”, “y”, “z” are presented in FIGURE 18.

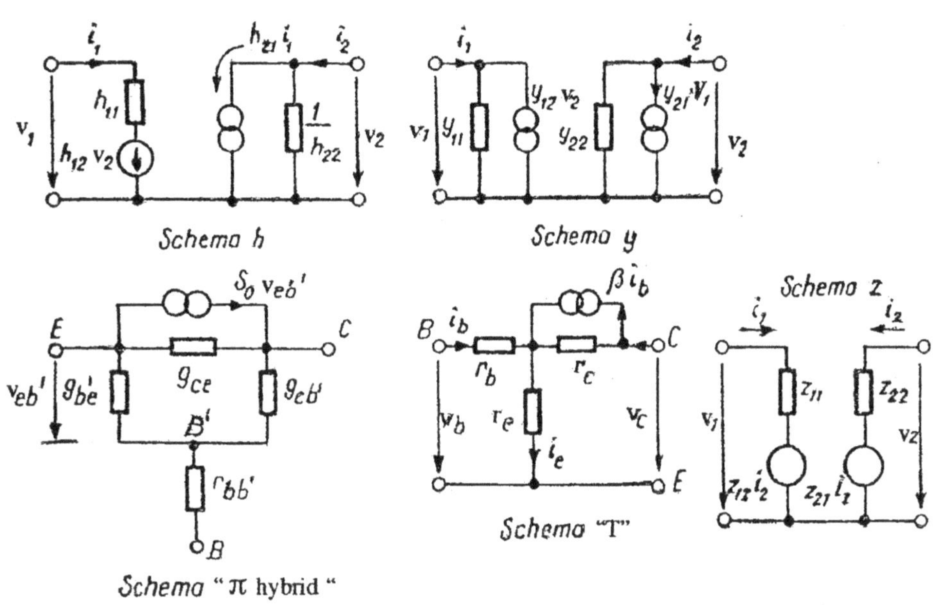

FIGURE 18 Equivalent circuits of the transistor

The equations of the quadripole using the “h” parameters are:

$$v_1 = h_{11}\, i_1 + h_{12}\, v_2$$
$$i_2 = h_{21}\, i_1 + h_{22}\, v_2$$

$h_{11} = v_1/ i_1$ for $v_2=0$ h_{11} is the input impedance with the output in short-circuit

$h_{12} = v_1/v_2$ for $i_1=0$ h_{12} is the reverse transfer factor in voltage with the input open circuit (reverse reaction factor)

$h_{21} = i_2/ i_1$ for $v_2=0$ h_{21} is the forward transfer factor in current (forward current gain) with the output in short circuit (å and ß in dynamic regime)

$h_{22} = i_2/v_2$ for $i_1=0$ h_{22} is the output admittance with the input open circuit

In *Table 2* are presented the approximated transition relations between the “h” parameters in common-base CB and in common-emitter CE configurations of the transistor.
The numerical values belong to the transistor BC109A for I_e=2mA V_{ce}=5V f=1kHz.

Table 2			
Transition Relations between "h" parameters of CB and CE configurations "h_b" and "h_e"			
CB			
h_{11b}	h_{12b}	h_{21b}	h_{22b}
$h_{11e}/(1+h_{21e})$	$h_{11e}h_{22e}/(1+h_{21e}) - h_{12e}$	$-h_{21e}/(1+h_{21e})$	$h_{22e}/(1+h_{21e})$
14.2 Ω	$1.55\ 10^{-4}$	-0.995	$10^{-7}\ \Omega^{-1}$
CE			
h_{11e}	h_{12e}	h_{21e}	h_{22e}
$h_{11b}/(1+h_{21b})$	$h_{11b}h_{22b}/(1+h_{21b})-h_{12b}$	$-h_{21b}/(1+h_{21b})$	$h_{22b}/(1+h_{21b})$
3 kΩ	10^{-4}	210	$0.85\ 10^{-5}\ \Omega^{-1}$
Numerical values for BC109A transistor			

In FIG 19 is presented the equivalent electrical circuit for an amplifier with transistor as an electrical quadripole with “h” parameters valid for any transistor configuration.
R_g is the electrical resistance of the signal generator and R_L is the load resistance.
Further are given relations with “h” parameters to calculate the properties of the equivalent electrical circuit:

- current gain $A_i = h_{21}/(1+h_{22} R_L)$
- voltage gain $A_v = -R_L h_{21}/(h_{11}+R_L \Delta h)$ $\Delta h = h_{11}h_{22} - h_{12} h_{21}$
- power gain $A_p = h_{21}^2 R_L/[(1+h_{22} R_L)(h_{11}+R_L \Delta h)]$
- input resistance $R_{input} = (h_{11}+R_L \Delta h)/(1+h_{22} R_L)$
- output resistance $R_{output} = (R_g+h_{11})/(h_{22} R_g+ \Delta h)$

gain = amplification

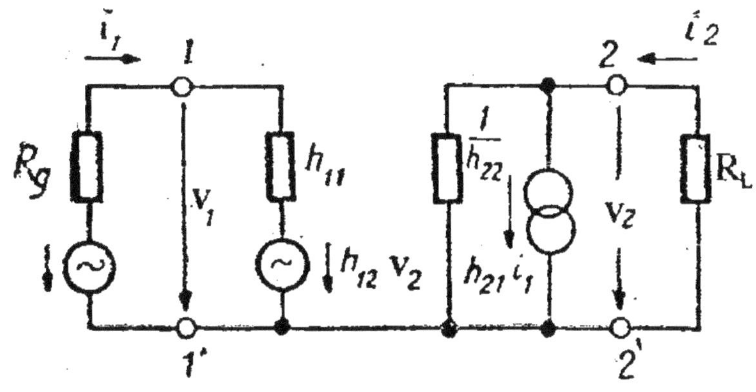

FIGURE 19 Equivalent circuit for amplifier with transistor as quadripole with "h" parameters

To characterise an amplifier with transistor it is necessary to calculate the current, voltage and power amplifications (gains) also the input and output resistances or impedances. The comparative presentation of the properties of the amplifier with transistor in the three basic configurations CB, CE, CC is given in *Table 3.*

Table 3

Transistor Configurations Properties

Parameter	CE	CB	CC
Input Resistance $R_{inp.}$	medium $R_{inp.\,e}$	low $R_{inp.\,b} = R_{inp.\,e}$ /ß	high $R_{inp.\,c} \approx$ ßR_L
Output Resistance $R_{out.}$	high $R_{out.\,e}$	very high $R_{out.\,b} \approx$ ß$R_{out.\,e}$	low $R_{out.\,c} = (R_{inp.\,e}+R_g)$/ß
Current Gain A_i	high ß	sub-unit å = ß /(ß+1)	high ¥ = ß + 1
Voltage Gain A_v	high	high	sub-unit
Power Gain A_p	very high	high	medium
Limit Frequency	low f_ß	high f_å = ßf_ß	low ≈ f_ß

Important for the transistor function is the *transconductance* defined as the ratio of the change in collector current to the change in base voltage over a small interval on the I_C -V_{BE} characteristic curve approaching the slope of a line tangent to the curve at a specific point. The transconductance is called also *mutual conductance* g_m or *slope S.*

$$S = g_m = \Delta I_C / \Delta V_{BE} \mid_{VCE=ct} \quad (mA/V)$$

For the bipolar junction transistor BJT the collector current is:

$$I_C = (1-\partial)\, I_E = (1-\partial)\, I_{EBO}\, [\exp (V_{EB} / V_T) - 1]$$

The *mutual conductance* g_m or *slope S* is:

$$g_m = \partial I_C / \partial V_{BE} \mid_{VCE=ct} = I_C / V_T$$

$g_m = 40\, I_C$ for Ge transistor at 17°C

From the physical viewpoint the most representative equivalent electrical circuits of the transistor at low frequency are "T" and "π hybrid" from FIGURE 18.

In the "T" circuit the resistances r_e r_b r_c are the internal resistances of the emitter, base and collector independent of the transistor configuration.

For the "π hybrid" circuit introducing the intrinsic slope S_0 important notations and relations are:

$S_0 = i_C / V_T = å\, i_E / V_T$ (mA/V)

$S_0 \approx 40\, i_C \approx (30 - 38)\, i_E$ for Ge transistor at 17°C

$S_0 v_{eb'}$ = emitter–collector current (generator)

$g_{b'e} = S_0 / ß$

$g_{ce} = S_0 / K$

$g_{b'c} = S_0 / (ß\, K)$

ß = base-collector current gain

K = internal reaction coefficient

$r_{bb'}$ = electrical resistance from the external base contact B to the intrinsic base B′

Based on a complete theoretical analysis of the transistor, it is possible to deduce for the transistor a natural equivalent circuit valid for any operating frequency.
That is the natural equivalent circuit "Giacoletto" presented in FIGURE 20.

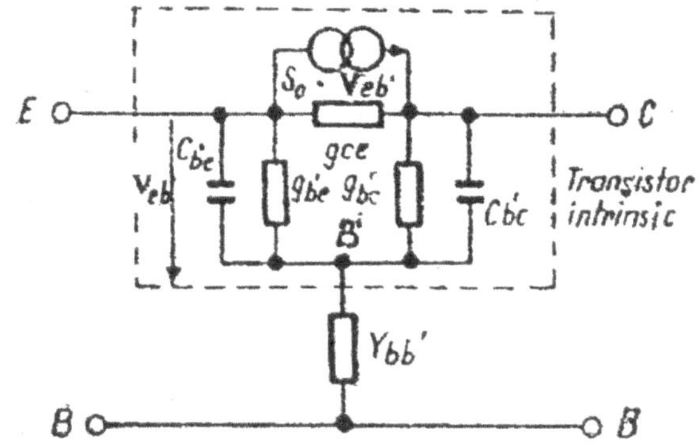

FIGURE 20 Equivalent circuit "Giacoletto"

The "Giacoletto" circuit is the "π hybrid" circuit for low frequencies plus the capacitances $C_{b'e}$ and $C_{b'c}$. Its advantage is its validity for a wide range of frequencies.
Important frequencies are:

$f_å$ frequency at which the current gain å in common-base CB configuration decreases at $1/\sqrt{2} = 0.707$ of its low frequency value

$f_ß$ frequency at which the current gain ß in common-emitter CE configuration decreases at $1/\sqrt{2} = 0.707$ of its low frequency value

f_T *transition frequency* at which the current gain ß drops at 1

Theoretically and experimentally the "Giacoletto" circuit is valid in the range:

$$f < f_å/4 \quad \text{or} \quad f < ßf_ß/3.3 \qquad \text{(Hz)}$$

Describing the amount of electrical noise introduced by transistor is the *noise factor F* defined as : F = output noise power / input noise power

or : $F = 10 \log_{10}$ (output noise power / input noise power)

For the noiseless transistor: F = 1 or F = 0 dB

COMMUTATION REGIME

In the commutation regime the transistor is in essence a controlled switch.
It is successively switched ON and OFF by a command signal applied on base, the transition between the two states being done at high speed.
The commutation regime is a reliable function from the viewpoint of the perturbations (voltage supply fluctuations, temperature, load etc).
The electronic commutation circuits can be easily combined with mechanical systems controlling them rapid and reliable.
In general in this regime the transistor is in the common-emitter configuration.
In the commutation regime the transistor works successively in the saturation state and in the cut-off state.
The simplest commutation circuit with transistor is the inverter presented in FIGURE 21.
A positive signal pulse at the transistor input is inverted into a negative signal pulse at its output.

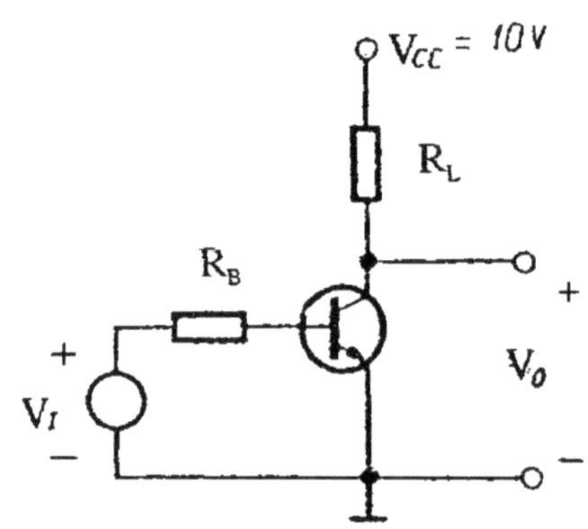

FIGURE 21 Inverter with npn transistor

The way the **npn** transistor in CE configuration commutes is illustrated on the output characteristic curve I_C-V_{CE} in FIGURE 22.
In the operating point T the transistor is in the cut-off state with both junctions in reverse conduction passed by their residual current I_R and the output voltage $V_0 = V_{CE} = V_{cc}$.

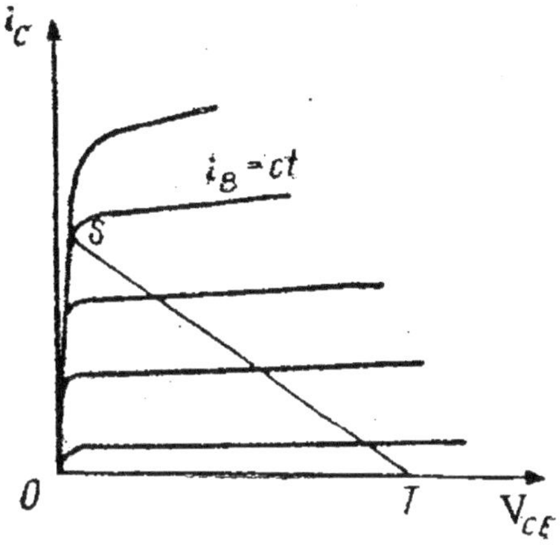

FIGURE 22 "Walk" of operating point in commutation

With a suitable positive command signal on base the transistor opens and commutes rapidly in the operating point S where it is in the saturation state.
In the saturation state the emitter junction and the collector junction are forward biased, the collector-emitter voltage is very low $V_{CE} = V_{CEsat}$ (it is 0,2–0.3 V for a Si transistor) and the collector current has the value:

$$I_{Csat} = (V_{cc} - V_{CEsat})/R_L \approx V_{cc}/R_L$$

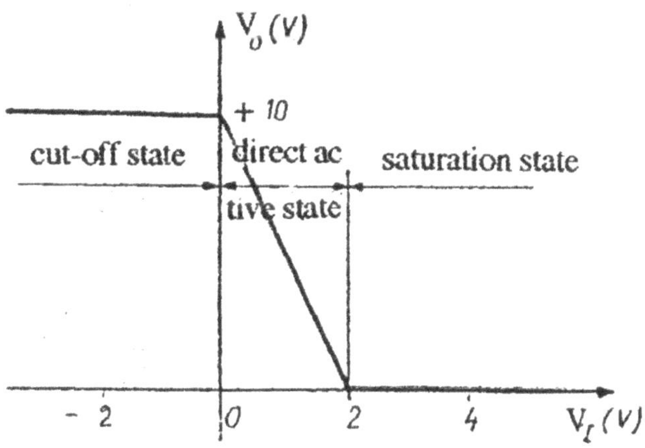

FIGURE 23 Output voltage / Input voltage characteristic for the inverter

To push the transistor in saturation regime, see FIGURE 23, the base current determined by the input voltage V_I should be:

$$I_B > I_C/\beta \approx V_{cc}/(\beta R_L) \qquad I_B = (V_I - V_{BE})/R_B$$

The electric charge stored in base is composed by the charge in normal active regime plus the excess charge in saturation regime.

The commutation times are the times necessary to pass from cut-off state to saturation state and inversely. In first approximation they correspond to the times to generate and evacuate the electric charge stored in base.
The time "ON" the transistor stays in saturation state is t_{on} and the time "OFF" the transistor stays in cut-off state is t_{off}.
The time necessary to "open" the transistor $t_{off\text{-}on}$ is mainly the rise time t_r : $t_{off\text{-}on} \approx t_r$
Rise time t_r = interval of time the collector current increases from 10% to 90% of its final maximum value I_{Csat}. The instantaneous collector current $i_c(t)$ and the collector-emitter voltage $v_{ce}(t)$, see FIGURE 24, are approximated to change in a linear fashion during the "open" switching interval:

$$i_c(t)=I_{Csat}\, t/t_r \qquad v_{ce}(t)=V_{cc}\,(1-t/t_r)$$

The time necessary to "close" the transistor $t_{on\text{-}off}$ is mainly the sum of the de-saturate time t_s and the fall time t_f: $t_{on\text{-}off} \approx t_s + t_f$
De-saturate time t_s = interval of time between the "close" command on base and the decrease of collector current at 90% of its maximum value. It is the most important since it is the main limitation of the commutation speed. Its value varies from 0.1μs to 10μs function of the transistor type.
Fall time t_f = interval of time the collector current decreases from 90% to 10% of its maximum value I_{Csat} and comparable with t_r. The instantaneous collector current $i_c(t)$ and the collector-emitter voltage $v_{ce}(t)$ are approximated to change in a linear fashion during the switching interval:

$$i_c(t)=I_{Csat}\,(1-t/t_f) \qquad v_{ce}(t)=V_{cc}\, t/t_f$$

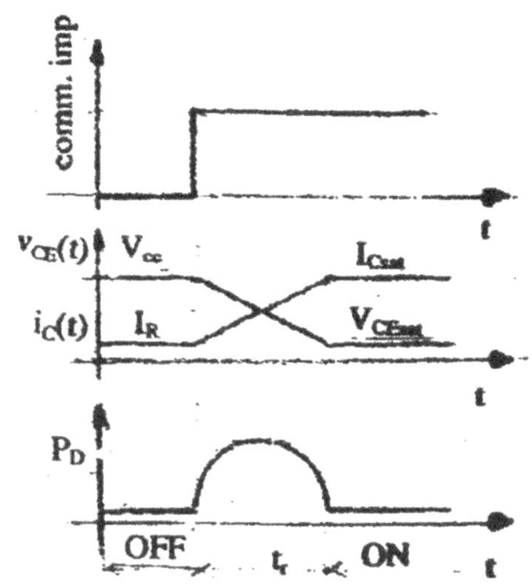

FIGURE 24 Inverter switching ON

The power dissipation on transistor: $P_D = I_C V_{CE}$

In the cut-off state: $P_D = I_R V_{cc}$

In the saturation state: $P_D = I_{Csat} V_{CEsat}$

In the transition time t_r: $P_D = (1/t_r) \int_0^{tr} i_c(t)\, v_{ce}(t)\, dt = I_{Csat} V_{cc}/6$

In the transition time t_f: $P_D = (1/t_f) \int_0^{tf} i_c(t)\, v_{ce}(t)\, dt = I_{Csat} V_{cc}/6$

The power dissipation on transistor and in consequence the transistor temperature is proportional with the *frequency of commutations f (Hz).*

More commutations per second, warmer the transistor.

The total power dissipation on transistor is:

$$P_D \approx f\,[I_R V_{cc}\, t_{off} + I_{Csat} V_{CEsat}\,(t_{on} + t_s) + I_{Csat} V_{cc}\, t_r/6 + I_{Csat} V_{cc}\, t_f/6] \qquad (W)$$

FUNCTIONAL LIMITS/ABSOLUTE LIMITS

The system of functional limits or absolute limits contains the limit values of some variables/parameters to assure an acceptable running of any device of a given type.

A. For any circuit containing semiconductor devices, the considerations regarding the temperature are vital.
In general can be defined three large areas of applications for transistors, each having among other features a specific range of ambient temperature T_A (referring to the cooling environment temperature):

Large consume applications	$T_A = \ \ 0 .... +55^\circ C$
Industrial applications	$T_A = -25 .... +75^\circ C$
Army applications	$T_A = -65 .... +125^\circ C$

For the operating junction it is considered a maximum possible temperature T_{jmax} that depends on the main material of the device.
For germanium (Ge) manufactured transistors $T_{jmax} = 85^\circ C$
For the silicon (Si) manufactured transistors $T_{jmax} = 175^\circ C$
The minimum operating temperature is for both types $T_{jmin} = -65^\circ C$
T_{jmax} and T_{jmin} values are absolute limits. Never should be exceeded.

For storage the limit temperature for the junction is higher.
For the germanium (Ge) manufactured transistors $T_{jmax\ storage} = 100^\circ C$
For the silicon (Si) manufactured transistors $T_{jmax\ storage} = 200^\circ C$
The minimum storage limit is for both transistor types $T_{jmin\ storage} = -65^\circ C$

$T_{jmax\ storage}$ and $T_{jmin\ storage}$ values are absolute limits. Never should be exceeded.

B. The collector current I_C is recommended by the manufacturer after statistical tests with satisfactory results from the viewpoint of the life time for a large number of devices.
The maximum value of the collector current I_{Cmax} is an absolute limit.
Never should be exceeded.

C. The power dissipation P_D in transistor leads to the heating of the junctions.

Its expression is: $P_D = V_{BE} I_E + V_{CB} I_C$

In normal active regime: $V_{BE} I_E << V_{CB} I_C$

In saturation regime: $V_{CB} I_C \approx V_{BE} I_E$

If their temperature increases the junctions can be damaged.

The maximum value of the power dissipation (direct current or average) P_{Dmax} is an absolute limit. Never should be exceeded.

D. The thermal resistance R_{th} is a technological parameter that represents the ratio of the difference between the junction temperature and that of a specified external reference point (ambient temperature) to the constant dissipated power in device in stationary regime P_D :

$$R_{th} = (T_j - T_A) / P_D \qquad (^\circ C/W)$$

Higher the thermal resistance more difficult is spread the heat of the junction in the environment with temperature T_A and so the permitted power dissipation is smaller:

$$P_{Dmax} = (T_{jmax} - T_A) / R_{th} \qquad (W)$$

However in FIGURE 25 it is shown that for the same R_{th} a silicon transistor (BFX 33) has a higher maximum power dissipation P_D than a germanium transistor (AC 117) because of the superior value of maximum operating temperature of junction T_{jmax}.

The equivalent circuit to cool a transistor is presented in FIGURE 26 where:

R_{thJC} is the thermal resistance between the junction and capsule (case)

R_{thCA} is the thermal resistance between the capsule and ambient

R_{thJA} is the thermal resistance between the junction and ambient

$$R_{thJA} = R_{thJC} + R_{thCA}$$

For the high power transistors R_{thJC} value is specified in catalogues. The designer can decrease R_{thCA} by placing the transistor on a heat sink, a cooling radiator with or without forced cooling; bigger the radiator surface smaller the resulting R_{thCA}.

For the low power devices the designer does not intervene in the thermal characteristics and in catalogues is specified R_{thJA}.

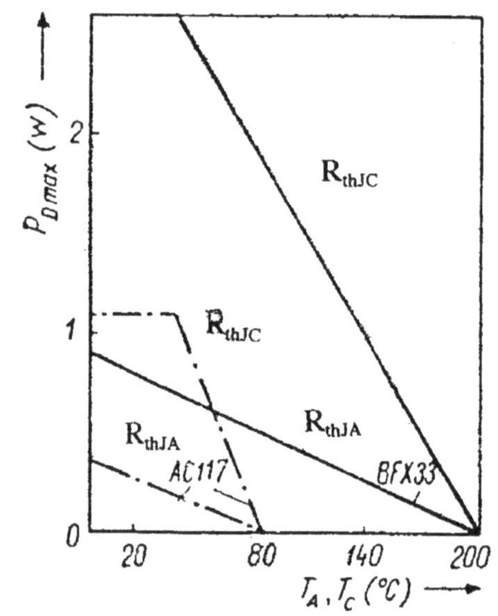

FIGURE 25 Maximum power dissipation function of ambient temperature and capsule temperature for Ge transistor AC117 and Si transistor BFY 33 with the same R_{th}

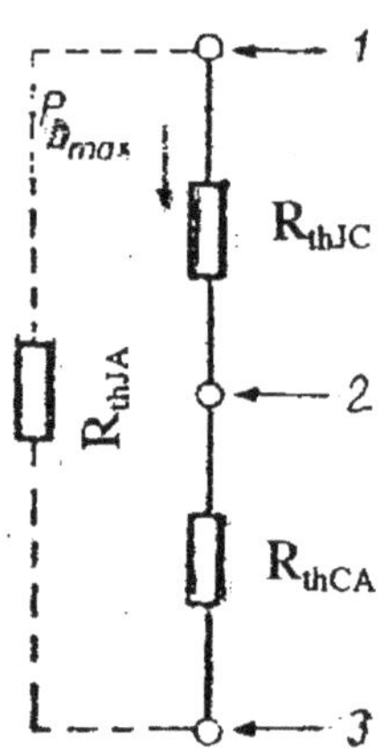

FIGURE 26 Equivalent cooling circuit of a transistor

Thermal design is indispensable when using the semiconductor for high power applications or using it under high operating temperature.
The concept of thermal resistance is used when considering heat dissipation.
The maximum value of the thermal resistance R_{thmax} is an absolute limit. Never should be exceeded.

E. At high reverse voltages V_R applied on the transistor junctions the mobile electric charge carriers, practically the electrons, collide and ionise neutral atoms.
Their multiplication gets the character of an avalanche and in consequence the current increases suddenly.
The voltage for which the current increases to infinity is called the breakdown voltage.
So another important limit in the transistor function in the normal active regime or in the cut-off regime is the maximum reverse voltage V_{Rmax} applied on junctions, which is also an absolute limit. Never should be exceeded.

UNIPOLAR TRANSISTOR

Representative for the unipolar transistor is the *Field Effect Transistor* FET.
FET is an active device whose output electric current is controlled by an electric field.
The two types of FET presented further are:
Field Effect Transistor with junction gate J-FET
Field Effect Transistor with insulated gate MOS-FET *with induced* or *initial channel*
The FET has specific advantages/possibilities in comparison with BJT as:

- Functions with only one type of electricity carriers and for that is much more stable with the temperature
- Assures a lower noise and distortions level up to very high frequencies
- Has good commutation characteristics
- Presents a very high input impedance
- Has high current gain and high power gain
- Is immune to radiation

As disadvantage is the lower voltage gain as consequence of the lower transconductance since the parabolic dependence of the output current on the input voltage $I_D \approx V_{GS}^2$.
Essentially, the field effect transistor combines the best properties of the bipolar transistor and the vacuum tube, and is incorporated easily in the integrated electronic circuits.
Its symbols and structure sketches are presented in TABLE.
J-FET has at the input a junction working in reverse conduction at a very low input current and MOS-FET has an insulated input (no junction), so the input impedance of J-FET is lower than that of MOS-FET.
As a consequence MOS-FET is more suitable in electrometric applications than J-FET.
For this reason and also because is manufactured easier MOS-FET is more spread.
But because parasitic input voltages can destroy the insulating thin oxide layer of MOS-FET input, on its structure it is built also a protection system with a Zener diode for $V_Z = 30\text{-}50V$ or with another MOS-FET with $V_{threshold} = 50\text{-}60V$.

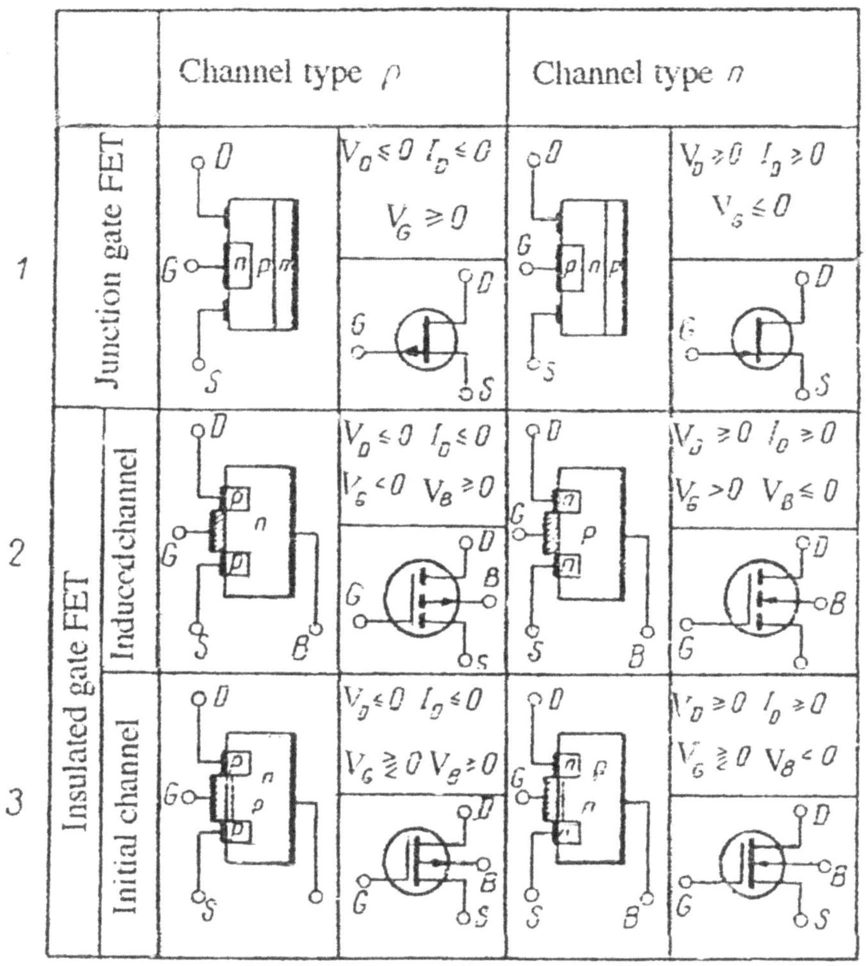

TABLE

Physical structure, symbol and sign for voltages and currents of FET in normal operating conditions. Source S is reference electrode.

FIELD EFFECT TRANSISTOR WITH JUNCTION GATE J-FET

The typical structure of the field effect transistor with junction gate J-FET is illustrated in FIGURE 27 where two spatial electric charge regions, corresponding to two **pn** junctions bound the **n** *conduction channel* of the current.

The two **p** type regions are connected together in the gate terminal G.

The ends of the channel are connected to other two terminals: source S and drain D.

FIGURE 27 illustrates the J-FET with **pnp** structure and **n** type channel.

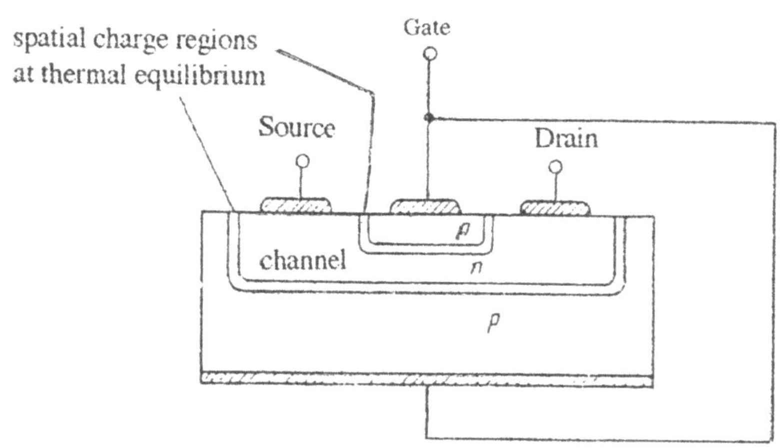

FIGURE 27 Transverse section through a J-FET with n channel

Applying a voltage between drain and source, a current of electrons which are negative electricity carriers will appear through the channel.

The source is always considered the reference terminal for the gate and drain voltages.

J-FET works usually with the two **pn** junctions in reverse conduction.

From the theory of the **pn** junction it is known that the region with spatial electric charge misses the mobile electrical charge carriers.

The control of the current is realised in this structure by the penetration of the spatial electric charge region in the channel (reducing so the channel width) by increasing the reverse bias of the junctions (increasing the negative voltage on gate).

Reducing the width of the channel is reduced the electrical conductance between the source and drain, the value of drain current decreases (the current of electrons from source to positive drain generated by the voltage V_{DS}).

The above description justifies the term "field effect" used in the name of the transistor. This term describes the control mechanism of the current as an effect of the variation of the channel conductance (by changing the width of the channel) function of the intensity of the electric field in the gate-channel zone, intensity determined by the gate voltage.

A normal function of J-FET implies the gate-channel junctions in reverse conduction. In the case of the **n** channel, for control, the gate should be electrically negative relative to source. A **npn** semiconductor structure will have a **p** channel and for control the gate should be electrically positive relative to source.

The dependence of the drain current on the drain voltage implies two cases a & b:

a. Low drain voltage V_{DS} (0.1-0.3V) where I_D varies linear with the V_{DS}

That region of I_D-V_{DS} characteristic curve is called the *linear region* (triode type).

J-FET behaves like an resistor with its resistance controlled by the gate-source voltage.

b. High drain voltage V_{DS} where I_D is independent of V_{DS} depending only on V_{GS}

That region of I_D-V_{DS} characteristic curve is called the *saturation region* (pentode type).

In the saturation region the dependence of the drain current I_D on the gate-source voltage V_{GS} is well approximated by the relation describing the transfer characteristic:

$$I_D = I_{D0} \, (1-V_{GS}/V_{thr})^2$$

I_{D0} drain current in saturation for $V_{GS} = 0V$

V_{thr} *threshold voltage* meaning the cutting/strangling gate-source voltage (at which the current ceases to be)

Practically in saturation the drain current varies with the drain voltage, because of the real value of the device output resistance, how can be seen in the experimental or typical I-V characteristic curves for J-FET with **n** channel presented in FIGURE 28. They are very alike with those of the pentode vacuum tube.

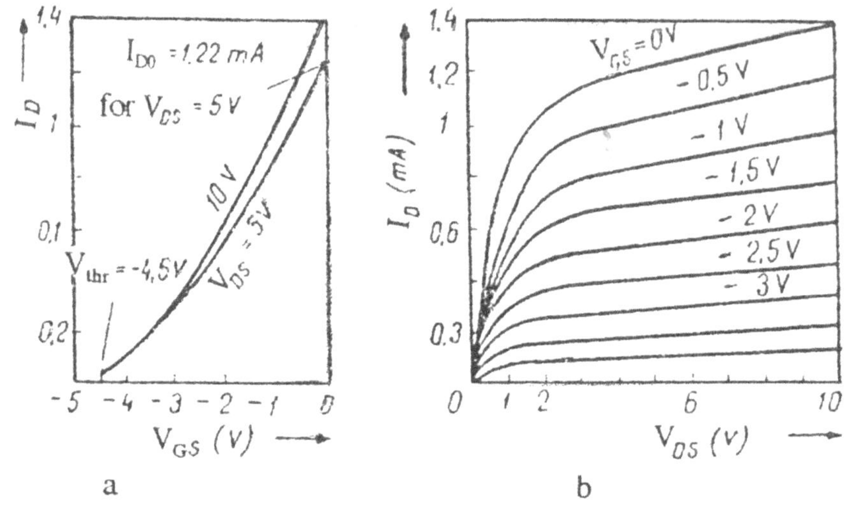

FIGURE 28 Typical I-V characteristics for low power, n channel J-FET
a. I_D-V_{GS} b. I_D-V_{DS}

These are the transfer characteristic curve $I_D = f(V_{GS})$ for V_{DS} = ct and the output (drain) characteristic curve $I_D = f(V_{DS})$ for $V_{GS} = ct \leq 0$.

In FIGURE 28a are given two transfer characteristic curves for V_{DS} = 10V / 5V.

In FIGURE 28b is given a family of output characteristic curves.

For the J-FET with **p** channel the characteristic curves have the same shape (are changed the direction of the currents and the sign of the voltages, the representation being in the third quadrant).

FIELD EFFECT TRANSISTOR WITH INSULATED GATE MOS-FET

Typically the structure of the field effect transistor with insulated gate MOS-FET involves an insulating layer between gate and its induced or initial channel.

MOS-FET WITH INDUCED CHANNEL

The *MOS-FET with induced channel* is a silicon monocrystal layer low contaminated with impurities Base in which are diffused two zones strong contaminated with the opposite type of impurities to those in base, Source and Drain, see FIGURE 29.
On the entire surface of the silicon layer is grown a layer of Silicon Dioxide; it will be removed where are located the drain and the source and there is deposited a thin layer of metal, forming the ohmic contacts named source S and drain D.
A thin layer of metal is deposited also on the oxide between source and drain forming the ohmic contact of the gate G and on the other side of the base forming the ohmic contact of the base B.
The source is always considered the reference terminal for the gate and drain voltages.
Typical parameters are:

channel length ≈ 0.3 μm
source and drain junction depth ≈ 0.2 μm
oxide thickness ≈ 10 nm
regions doping ≈ 10^{20} cm^{-3}
substrate doping ≈ $3 \cdot 10^{17}$ cm^{-3}

When the gate terminal is not under voltage, even if there is a drain-source voltage V_{DS}, will not appear a drain current I_D, because the source zone and the drain zone together with the semiconductor between them build up two **pn** junctions connected in opposition so equal of the polarity of the voltage between the source and drain one junction is reverse biased, stopping the conduction path between source and drain through the intrinsic semiconductor (cut-off regime).

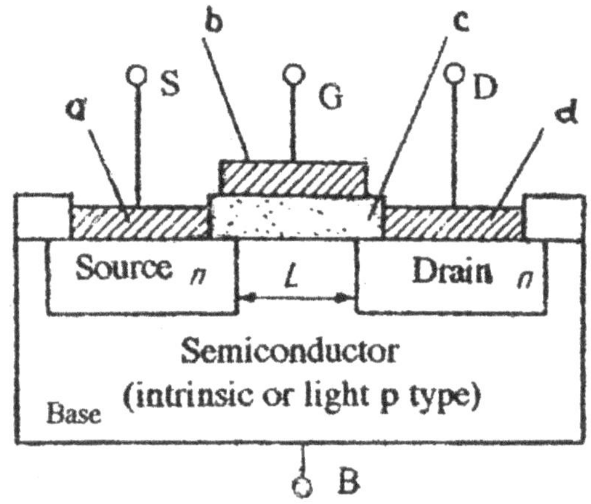

FIGURE 29 Physical structure of a MOS-FET in a transverse section
a. metallic contact on source region b. metallic contact on oxide layer
c. oxide layer d. metallic contact on drain region

In FIGURE 30a it is shown that when the gate is connected to a positive voltage relative to source and drain appears an electric field which facilitates the accumulation of a negative charge under the gate (inversion layer), electrons extracted from the source and drain regions, which formes a channel called *induced channel* permitting a current between source and drain. That is the *Field Effect Transistor with insulated gate and induced channel* or *MOS-FET with induced channel.*

The electrical conductance of the channel depends on the potential difference between gate and source.

The previous description justifies the term "field effect" used in the name of the transistor. The term describes the control mechanism of the drain current as an effect of the variation of the channel conductance function of the intensity of the electric field in the channel zone, intensity determined by the gate voltage.

In FIGURE 30b is presented the characteristic curve $G_c = f(V_{GS})$ where G_c is the induced channel conductance, V_{GS} is the voltage drop between gate and source and V_{thr} is the threshold voltage.

The drain current I_D that passes the channel is the consequence of the drain-source voltage drop V_{DS} and is controlled by the gate-source voltage drop V_{GS}.

Its dependence on the drain voltage implies two cases a & b:

a. Lower drain voltages $V_{DS} < V_{GS}$ where I_D varies linear with V_{DS}

In FIGURE 30c are presented the output (drain) characteristic curves $I_D = f(V_{DS})$ in the *linear region*, for low drain voltages $V_{DS} < V_{GS}$ for 2 values of V_{GS} : +1V and +2V. In consequence in the linear region MOS-FET can be used as voltage controlled resistance.

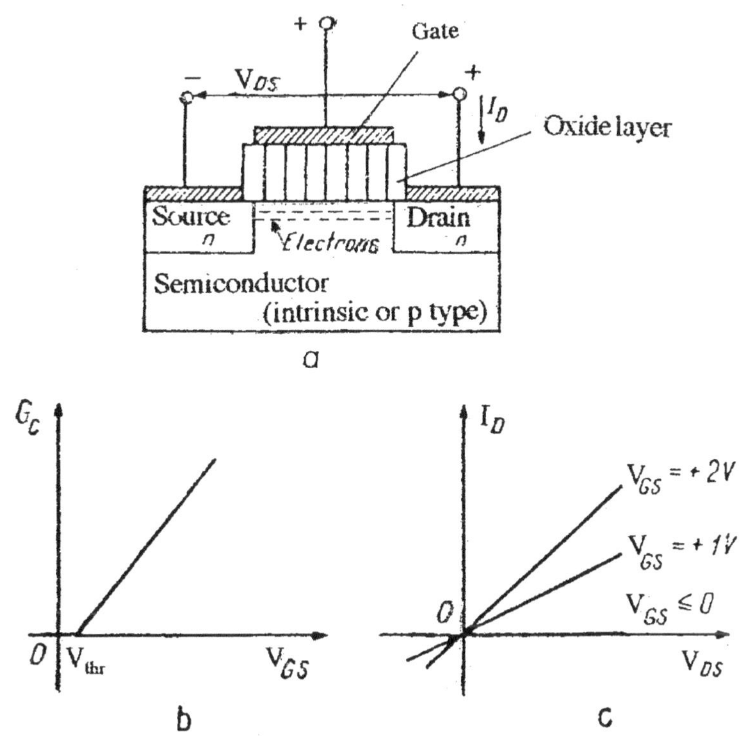

FIGURE 30
Inducing a conductive channel between source and drain
a. physical situation
b. conductance between source and drain
c. I_D-V_{DS} characteristics for low V_{DS}

b. Higher drain voltages $V_{DS} \geq V_{GS}$

If $V_{DS} \approx V_{GS}$ the intensity of the electric field varies along the channel and the output (drain) characteristic curves become non-linear in the *saturation region*, see FIGURE 31. The drain current becomes independent of drain voltage and enters the saturation regime where is well approximated by the relation describing the transfer characteristic:

$$I_D = (\beta/2)(V_{GS} - V_{thr})^n$$

n is close to 2 and ß is a manufacturing parameter approximately $0.8 mA/V^2$

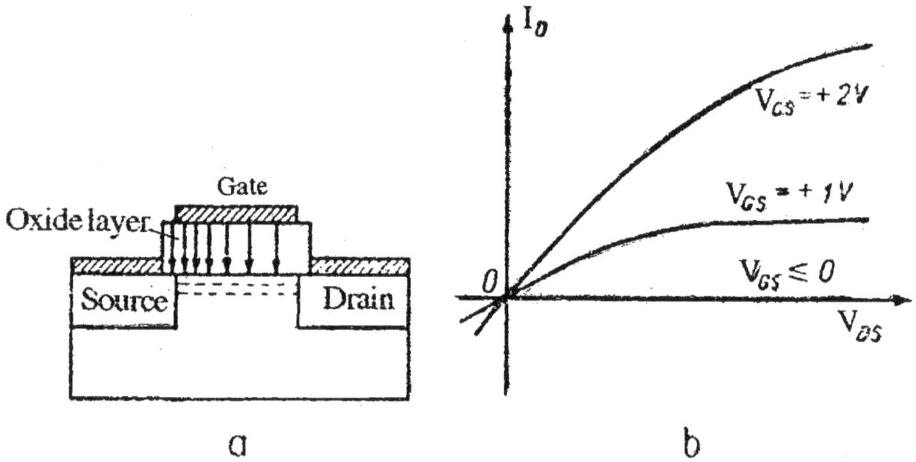

FIGURE 31
When the drain-source voltage is not negligible the drain characteristics become non-linear: a. physical situation
b. I_D-V_{DS} characteristics

In FIGURE 32 are presented the complete output (drain) characteristic curves $I_D = f(V_{DS})$ for MOS-FET with induced channel for three values of V_{GS} : +5V, +10V, +15V
The experimental output (drain) characteristic curves in FIGURE 33 are different from the theoretical ones because in the saturation region they have a slope different of zero with the increase of the V_{DS} (channel shunting).

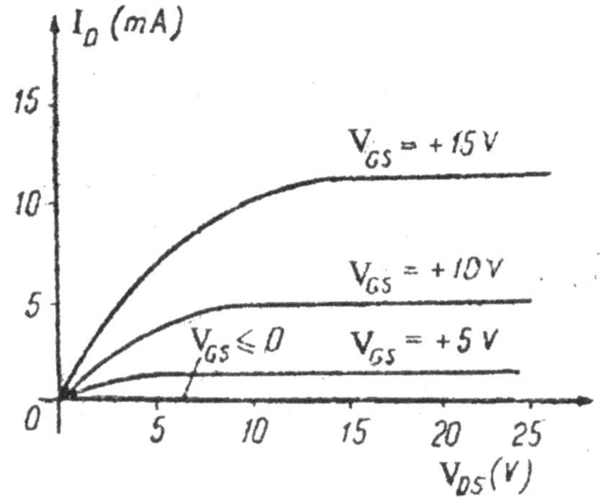

FIGURE 32 Complete I_D-V_{DS} characteristics for MOS-FET with induced channel; for $V_{DS} > V_{GS}$ the current saturates

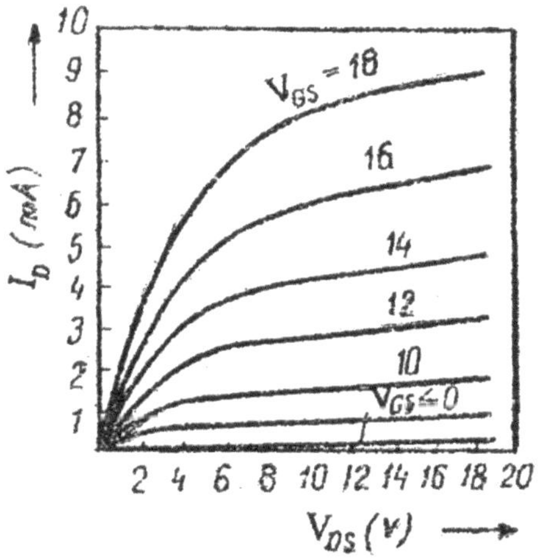

FIGURE 33 Experimental I_D-V_{DS} characteristics for MOS-FET with induced channel

Using a similar procedure it is possible to manufacture a MOS-FET with **p** induced channel. The source and drain regions are strong **p** contaminated (doped) and work with negative voltage on drain and gate.

Its characteristics are represented in the third quadrant.

MOS-FET WITH INITIAL CHANNEL

There are also field effect transistors with insulated gate which are built with channel whose conductivity or number of mobile electric charge carriers can be increased or decreased applying either a positive or a negative voltage between gate and source V_{GS}. $V_{GS} \leq 0$ and/or $V_{GS} \geq 0$ control the conductance of its initial channel. That is the *Field Effect Transistor with insulated gate and initial channel* or *MOS-FET with initial channel.*
The family of the experimental output (drain) characteristic curves $I_D = f(V_{DS}, V_{GS})$ for the MOS-FET with initial channel is presented in FIGURE 34.

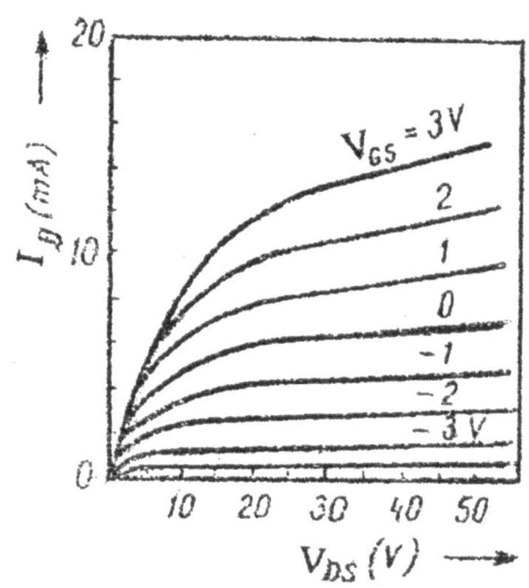

FIGURE 34 Experimental I_D-V_{DS} characteristics for MOS-FET with initial n channel

In general the source is considered the reference electrode for voltages in what case it is said that FETs are used in common source configuration CS.
However they can be used also in common gate configuration CG or common drain configuration CD (source follower) as in FIGURE 35.

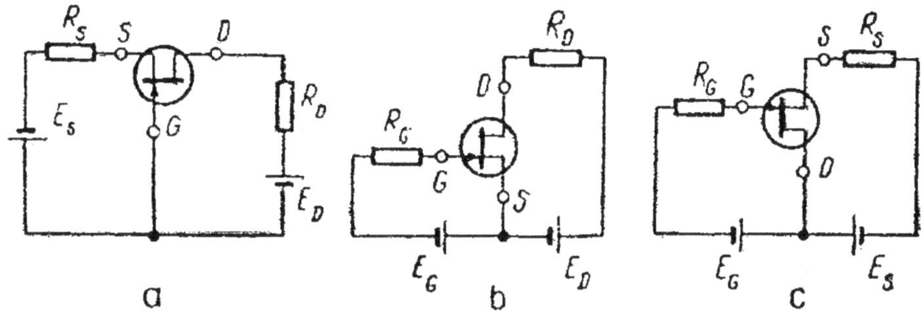

FIGURE 35 FET Configurations a. common gate CG
b. common source CS
c. common drain CD

The field effect transistors have functional limits for the values of the drain-source voltage V_{DS}, the gate-source voltage V_{GS}, the drain current I_D and the dissipated power on transistor P_D which never should be exceeded; they are found in catalogues.

In catalogues are presented too the built-in parameters: I_{D0} V_{thr} $r_{DS}(ON)$ g_{m0}

STATIC REGIME

When a FET operates at certain values of I_D, V_{DS}, V_{GS}, it operates in the static or stationary regime in a certain *operating point*.

The operating point is realised connecting between the terminals of the transistor resistors with suitable values, resulted from design, to bias the transistor.

The operating point is chosen on the static output (drain) characteristic curve $I_D = f(V_{DS}, V_{GS})$ and on the static transfer characteristic curve $I_D = f(V_{GS})$ how can be seen in FIGURE 36 the point $A(I_{DQ}\ V_{DSQ}\ V_{GSQ})$.

In amplifiers, FET is used almost always in the saturation regime of the drain current (for what reason further I_D will mean I_{Dsat}) for which the static transfer characteristic $I_D = f(V_{GS})$ can be expressed for a J-FET or a MOS-FET with initial channel:

$$I_D = I_{D0}\,(1 - V_{GS}/V_{thr})^2$$

For MOS-FET with **n** type induced channel working in the saturation regime at a constant drain-source voltage, the static transfer characteristic $I_D = f(V_{GS})$ can be expressed:

$$I_D = (ß/2)(V_{GS} - V_{thr})^2$$

The minimum drift with the temperature of the operating point (drift caused by the intrinsic conduction factor and by the threshold voltage) can be obtained choosing as operating point the point Z on the characteristic curve presented in FIGURE 36c that has no variation with the temperature.

For J-FETs the coordinates of the point Z ($V_{GSZ}\ I_{DZ}$) are:

$$|V_{GSZ}| \approx |V_{thr}| - 0.63 \qquad \text{and} \qquad I_{DZ} = I_{D0}\,(0.63/V_{thr})^2$$

But this point is not sufficiently reproducible for all FETs of the same manufacturing because V_{thr} and I_{D0} can't be the same.

For MOS-FETs is even more difficult to obtain the point where the drift is null.

For this reason are preferred bias circuits to stabilize the drain current by negative reaction of the direct current.

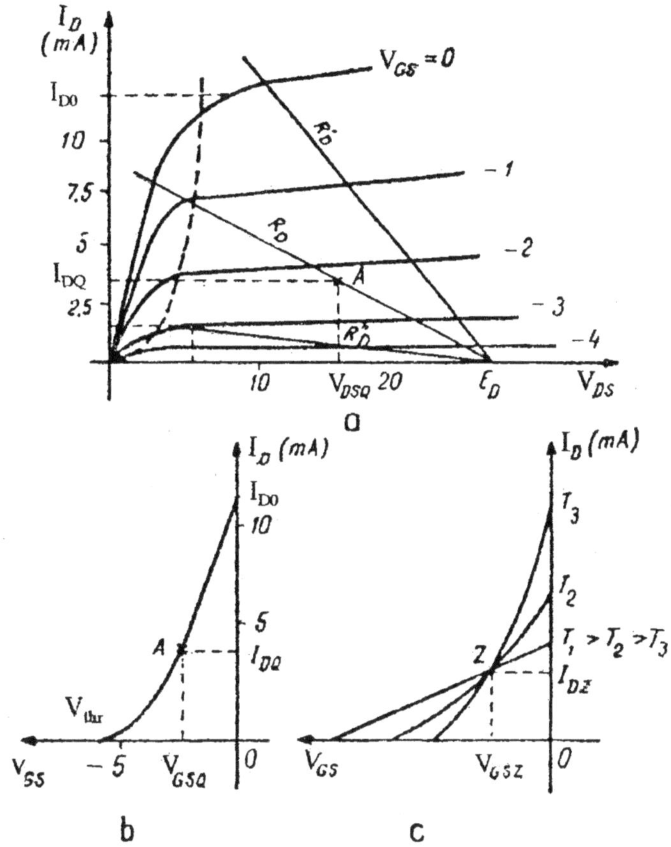

FIGURE 36 Choosing the operating point of FET

a. in I_D-V_{DS} plane
b. in I_D-V_{GS} plane
c. in I_D-V_{GS} plane at a drain current independent of temperature

The simplest bias circuit for J-FET or MOS-FET with initial channel is similar to that used for vacuum tubes, which stabilises the operating point by the negative reaction introduced by R_S, see FIGURE 37a.

To obtain the wanted operating point, I_{DQ} V_{DSQ} V_{GSQ}, the resistors R_S and R_D should have the values $R_S = -V_{GSQ}/I_{DQ}$ $R_D = (E_D - V_{DSQ} - R_S I_{DQ})/I_{DQ}$

The circuit assures also the stability of the operating point against manufacturing drifts.

For a better stabilization of the operation point it is used the circuit from FIGURE 37b with a voltage divider at input.

In this case the resistor R_S should have the value: $R_S = (V_G - V_{GSQ})/I_{DQ}$

For R_{G1}, R_{G2} are chosen not too high values to have V_G well defined.

To bias MOS-FET with induced channel it is not possible to use the circuit from FIGURE 37a because a voltage drop on R_S biases reversely the gate.

It is possible to use the circuit from FIGURE 37b with the condition that $|V_G| > |V_{thr}|$ assuring a suitable bias of gate.

It is possible to use also the bias circuit from FIG 37c which assures more linear function with undistorted signal for $V_{GS} < V_{DS}$ stabilizing the operating point by negative feedback introduced by R_{G1} R_{G2} but in the same time the gain and input impedance are low.

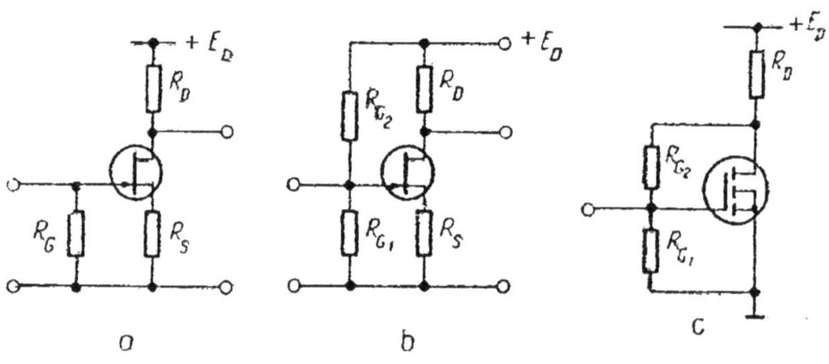

FIGURE 37 Bias circuits

a. automatic bias circuit of gate for J-FET or MOS-FET with initial channel
b. bias circuit of gate with voltage divider
c. voltage divider connected to drain to bias the gate of MOS-FET with induced channel

DYNAMIC REGIME

When leaves the static regime and operates in the dynamic regime as low frequency amplifier, the field effect transistor has few important advantages over the bipolar transistor as high input resistance, high current gain, high power gain, better thermal stability, less distortions.

As disadvantage in competing with the bipolar transistor is mentioned its lower voltage gain.

The equivalent circuit of a FET as low frequency, low amplitude signal amplifier deduced from its real physical behaviour is presented in FIGURE 38.

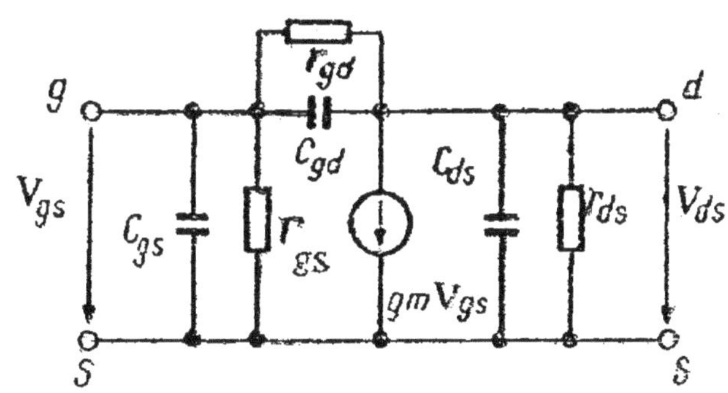

FIGURE 38   Equivalent circuit of FET at low frequency

As values can be admitted:

C_{gs} = 1...10pF C_{gd} = 0.5...5pF g_m = 0.5...5mA/V r_{ds} = 10...100KΩ

The value of the drain-source resistance r_{ds} is in general lower than the value of the collector-emitter resistance r_{ce} in the bipolar transistor contributing negatively to the voltage gain.

The gate-source electrical resistance r_{gs} value is much higher than the signal sources resistance or than the bias circuit resistances; it is taken in consideration only for the electrometric amplifiers or for J-FET working at high temperatures over 100 Celsius degrees.

The dependence of the drain current I_D by the gate-source voltage V_{GS} is determined by the *transconductance* defined as the ratio of the change in drain current to the change in gate voltage over a small interval on the I_D-V_{GS} characteristic curve approaching the slope of a line tangent to the curve at a specific point.
The transconductance is called also *mutual conductance* g_m.

$$g_m = \Delta I_D / \Delta V_{GS} \mid_{V_{DS}=ct} \qquad \text{(mA/V)}$$

In amplifiers, FET is used almost always in the saturation regime of the drain current for what reason further g_m will mean g_{msat}.

For J-FET or MOS-FET with initial channel the drain current in the saturation regime is:

$$I_D = I_{D0}\,(1 - V_{GS}/V_{thr})^2$$

and

$$g_m = \partial I_D / \partial V_{GS} \mid_{V_{DS}=ct} = -2I_{D0}\,(1 - V_{GS}/V_{thr})/V_{thr} = g_{m0}\,(1 - V_{GS}/V_{thr})$$

$g_{m0} = -2I_{D0}/V_{thr}$ is the value of the mutual conductance or transconductance for $V_{GS} = 0$

Those relations are verified experimentally very well for all FETs.
Considering the typical values I_{D0}=12.5mA and $V_{thr} = -5$V will result:
for $V_{GS} = 0$ $g_{m0} = 5$mA/V and
for $V_{GS} = -2.75$V $I_D = 2.5$mA $g_m = 2.25$mA/V
For J-FETs of different types $g_m \approx$ 0.1...10 mA/V.

For MOS-FET with induced channel the drain current in the saturation regime is:

$$I_D = (\beta/2)(V_{GS} - V_{thr})^2$$

and

$$g_m = \partial I_D / \partial V_{GS} \mid_{V_{DS}=ct} = \beta\,(V_{GS} - V_{thr}) = 2I_D/(V_{GS} - V_{thr})$$

ß is a built-in parameter for MOS-FET
Considering as typical value $\beta \approx 0.8$mA/V^2 for I_D=2.5mA results $g_m \approx 2$mA/V.
For MOS-FETs used in amplifiers for currents lower than 10mA $g_m \approx$ 0.1…5mA/V.
For comparison, the transconductance or mutual conductance of the bipolar transistor at same value of the current I_C=2.5mA is $g_m \approx 40I_C$ =100mA/V, value 40-50 times higher.
Because of its lower transconductance a FET has a lower voltage gain than a BJT.

To describe FET as high frequency amplifier it is used the equivalent circuit with the admittance parameters Y defined for the three working configurations:
common-source CS, common-gate CG, common-drain CD.
The Y parameters are complex and vary with the frequency and above 100-300MHz are difficult to be measured. At those frequencies the transistor is represented by a circuit with distributed parameters S (scattering parameters), which can be measured easier and from which are deducted the Y parameters.

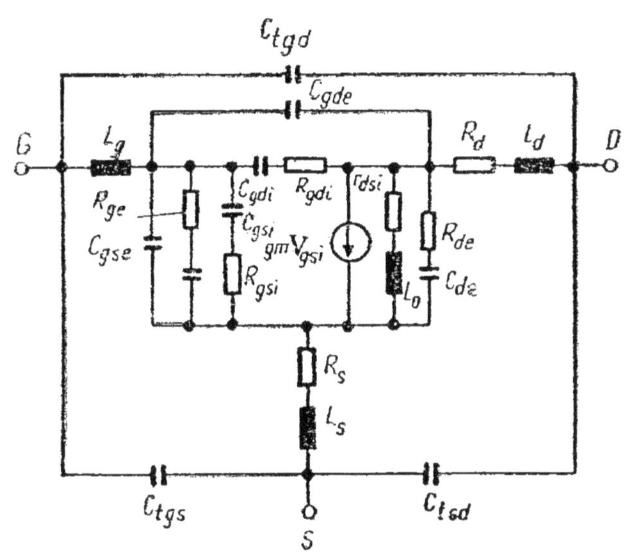

FIGURE 39 Complete equivalent circuit of FET for high frequencies deduced from physical parameters

The complete equivalent circuit for J-FET and MOS-FET deduced from physical parameters in the most general case is presented in FIGURE 39.
The frequency at which the module of the transconductance or mutual conductance g_m falls 3dB from its value at low frequencies is called the *cut-off frequency* f_0 and is defined by the relation: $f_0 = g_m/(2\pi C_g)$ (Hz)
C_g is the gate capacitance. f_0 has values of hundreds of thousands MHz for the intrinsic transistor but these values are not obtained experimentally for serial devices, meaning that the external elements as capacitors, source resistance determine the frequency performances.

VOLTAGE CONTROLLED RESISTANCE REGIME

The field effect transistor with junction J-FET and the field effect transistor with insulated gate MOS-FET around the origin at low values of the drain-source voltages V_{DS}, present an approximately linear dependence of the drain current I_D on the V_{DS} voltage.

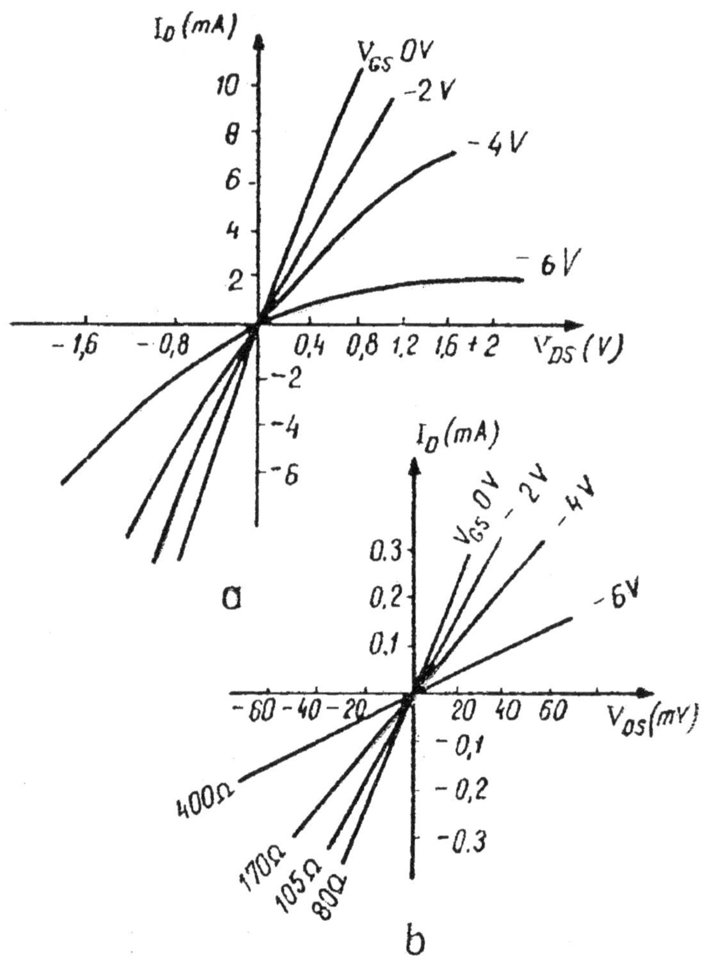

FIGURE 40 I_D-V_{DS} characteristics of FET as variable resistance
a. normal scale b. around origin

In consequence at low values of V_{DS}, the field effect transistor behaves like a resistance whose value is controlled by the gate-source voltage V_{GS}.

That can be illustrated by the family of characteristic curves $I_D(V_{DS})$ of a J-FET with **n** channel around the origin at two different scales in FIGURE 40.

For voltages V_{DS} between 100mV and 2 V the characteristic curves are not linear neither symmetrical: $I_D(V_{DS}) \# -I_D(-V_{DS})$ see FIGURE 40a.

But for voltages V_{DS} less than 100mV FET behaves like a conductance linear variable with V_{GS} of value g_{DS} equal to that of the transconductance or mutual conductance g_{msat} at the same value of V_{GS}:

$$g_{DS} = g_0 (1 - V_{GS} / V_{thr}) \qquad g_0 = 2I_{D0} / |V_{thr}| \qquad (\Omega^{-1})$$

$$r_{DS} = 1/g_{DS}$$

$$r_{DS} = r_0 /(1 - V_{GS} / V_{thr}) \qquad r_0 = |V_{thr}| /(2I_{D0}) \qquad (\Omega)$$

The minimum value of the J-FET resistance in conduction r_0 is obtained for $V_{GS} = 0$ when $I_D = I_{D0}$ and is of the order 2-300 Ω.

For MIS-FET the minimum values of the resistance that can be obtained are 20...1000 Ω.

The maximum values of the resistance which can be still controlled obtained before the strangling of the channel are in the order of 0.1...1 MΩ.

A cut-off FET presents between drain and source a resistance $r_{DS}(OFF)$ in the order of 10^8-10^{12} Ω.

Alike the resistance of a cut-off junction, $r_{DS}(OFF)$ decreases exponentially with the temperature, decreasing approximately 10 times when the temperature increases with 30° C.

COMMUTATION REGIME

In the commutation regime the field effect transistor switches from the cut-off state to the conduction state and vice versa at high speed with $t_{off\text{-}on}/t_{on\text{-}off}$ in ns range.

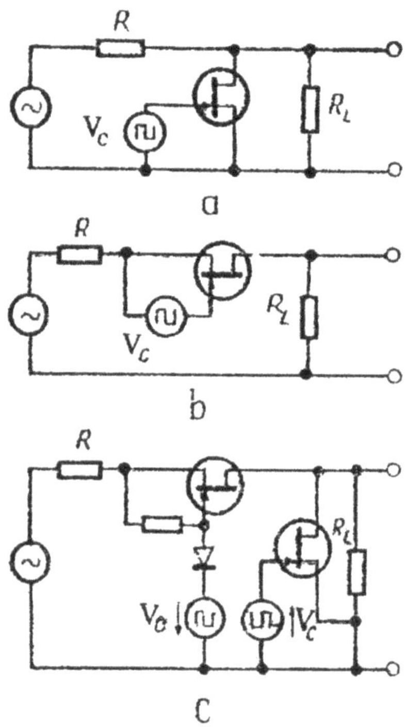

FIGURE 41 Commutation circuits with FET a. shunt b. series c. series-shunt

The field effect transistor is used frequently in the analogue commutation circuits where it is necessary an electric signal to be transmitted or stopped to be transmitted further function of a command.

The most used commutation circuits are: shunt, series, series-shunt see FIGURE 41.

In the shunt commutation circuit presented in FIGURE 41a the electrical signal is transmitted when the transistor is OFF; when the transistor is ON the signal is short-circuited and can not appear at the output. In the series commutation circuit presented in FIGURE 41b the electrical signal is transmitted at output only when the transistor is ON; when the transistor is OFF the signal is stopped to appear at the output.

The series-shunt commutation circuit presented in FIGURE 41c although is more complex than the previous two combines some of their advantages and reduces the errors. The series-shunt commutation circuit is used in the cutting up modulators for high precision where they have a low drift in time or with the temperature (0.1-1μV), having superior performances to those realised with bipolar transistors.

Table 4

Parameters of Typical Commutation FETs

Type	r_{DS} (ON) (Ω)	r_{DS} (OFF) (Ω)	I_{DS} (OFF) (nA)	I_{DG} (OFF) (nA)	V_{thr} (V)	C_{gd} (pF)	C_{gs} (pF)	C_{ds} (pF)
J-FET								
CMX 740	< 2.5		< 10	< 10	8	45	45	
CAG -6	< 6		0.5	0.5		25	25	
F.E. 0655A	< 25			< 2	- 10	12	12	
BSV -78	< 25	$>10^{11}$	< 0.25	< 0.25	- 7	5	5	
BFW -10	< 300		< 0.5	< 0.5	- 6	0.6	4	
MOS-FET								
HDSW -2106	20					11	11	8
MEM -563	50		< 1	0.01	1.5	0.3	4	3.5
BSV 81	< 50	$>10^{11}$	< 1	< 0.01	- 2	< 0.5	< 5	< 1.2
3N 150	100	$>10^{10}$	0.2	0.03	- 3.6	1.6	1.6	
MEM -520	150		0.2	< 0.03	- 4	1.5	2.5	0.15
3N 138	240	$>10^{10}$	0.01	1E-04	- 5	0.18	3	

Grace to their high speed in commutation the field effect transistors are used in the integrated technic.

As element of analogue commutation, the field effect transistor has the same static characteristics to those as voltage controlled resistance.

But in commutation are used the extreme characteristics corresponding to the minimum possible value of the resistance in open state $r_{DS}(ON)_{min}$ and the maximum possible value of the resistance in the cut-off state $r_{DS}(OFF)_{max}$.

$$r_{DS}(ON)_{min}=2...1000\ \Omega \qquad r_{DS}(OFF)_{max}=10^8...10^{12}\ \Omega$$

Values of r_{DS} for commutation field effect transistors are given in *Table 4*.

COMPARISON AND PRACTICAL APPLICATIONS

In *Table 5* is presented a comparison between the properties of the bipolar transistor and the properties of the unipolar transistor represented by J-FET and MOS-FET to attract the attention on their possible/typical utilizations.

Table 5

Comparison of Transistors Properties

Comparing	*Bipolar Transistor* - CE Configuration	*Unipolar Transistor* - CS Configuration	
		J-FET	MOS-FET
Input Resistance	Low 0.1 - 10 kΩ	High 0.1 - 1 GΩ	Very High 1 GΩ - 100 TΩ
Temperature Influence	Appears thermal inciting phenomenon	No thermal inciting phenomenon by adequate operating point	
Slope - low power transistors	20 - 200 mA/V	0.2 - 10 mA/V	
Current Gain - low frequency	20 - 60 dB	100 - 140 dB	140 - 180 dB
Voltage Gain - low frequency	20 - 50 dB	20 - 40 dB	
Maximum Frequency	100 - 1000 MHz	100 MHz	500 MHz

In *Table 6* is presented a comparison between the transistor types J-FET and MOS-FET regarding their most used fields of application. But in the same field is not excluded the presence of the other transistor type.

Table 6

Application Areas for J-FET and MOS-FET

	J-FET	**MOS-FET**
Commutation circuits	X	X
Switches	X	X
General use amplifiers	X	
Low noise, low frequency amplifiers	X	
High frequency amplifiers		X
Differential amplifiers	X	X
High input resistance (10^6-$10^9\Omega$) amplifiers	X	
Very high input resistance ($>10^{10}\Omega$) electrometer amplifiers		X
Low drift amplifiers	X	

CATALOGUE DATA

Data given for orientation

GENERAL USE SILICON DIODES										
Type	Absolute Limits				Electrical Characteristics					Case
	V_R(V)	I_F(mA)	P_D(mW)	T_j(°C)	V_F(V)	I_F(mA)	I_R(nA)	V_R(V)	R_{th}(°C/mW)	
BA170	20	150	300	150	1	80	50	10	0.41	DO35
BA171	30	150	300	150	1	80	50	15	0.41	DO35
BA172	50	150	300	150	1	80	50	25	0.41	DO35

SILICON RECTIFIER DIODES (ORDINARY AND FAST)									
Type	I_F(A)/T_A(°C)	I_0(A)/T_A(°C)	$V_{RRM}=V_R$(V)	I_{FSM}(A) 10ms	I_{FRM}(A)	V_F(V)/I_F(A)	I_R(mA)/T_A(°C)	t_{rr}(ns)	Case
0.4A									
BA157	0.5/25	0.4/25	400	15	2	1.5/0.4	0.005/25	200	F126
BA158	0.5/25	0.4/25	600	15	2	1.5/0.4	0.005/25	300	F126
BA159	0.5/25	0.4/25	1000	15	2	1.5/0.4	0.005/25	300	F126
BAX157	0.5/25	0.4/25	400	15	2	0.97/0.4	0.005/25	400	F126
DRR104	0.5/25	0.4/25	100	15	2	1.5/0.4	0.1/100	1000	F126
DRR204	0.5/25	0.4/25	200	15	2	1.5/0.4	0.1/100	1000	F126
DRR404	0.5/25	0.4/25	400	15	2	1.5/0.4	0.1/100	1000	F126
DRR604	0.5/25	0.4/25	600	15	2	1.5/0.4	0.1/100	1000	F126
DRR114	0.5/25	0.4/25	1000	15	2	1.5/0.4	0.1/100	1000	F126
0.75A									
F057	1/25	0.75/25	50	30	4	1/0.75	0.05/100		DO13
F087	1/25	0.75/25	80	30	4	1/0.75	0.05/100		DO13
F107	1/25	0.75/25	100	30	4	1/0.75	0.05/100		DO13
F207	1/25	0.75/25	400	30	4	1/0.75	0.05/100		DO13
F307	1/25	0.75/25	600	30	4	1/0.75	0.05/100		DO13
F407	1/25	0.75/25	800	30	4	1/0.75	0.05/100		DO13
1A									
1N4001	1.15/75	1/75	50	30	10	1.1/1	0.05/100		F126
1N4002	1.15/75	1/75	100	30	10	1.1/1	0.05/100		F126
1N4003	1.15/75	1/75	200	30	10	1.1/1	0.05/100		F126
1N4004	1.15/75	1/75	400	30	10	1.1/1	0.05/100		F126
1N4005	1.15/75	1/75	600	30	10	1.1/1	0.05/100		F126
1N4006	1.15/75	1/75	800	30	10	1.1/1	0.05/100		F126
1N4007	1.15/75	1/75	1000	30	10	1.1/1	0.05/100		F126
D1N13	1.15/75	1/75	1300	30	10	1.1/1	0.05/100		F126
D1N16	1.15/75	1/75	1600	30	10	1.1/1	0.05/100		F126
2A									
F102	2.4/25	2/25	100	70	6.5	1.2/2	1/150		DO13
F202	2.4/25	2/25	200	70	6.5	1.2/2	1/150		DO13
F402	2.4/25	2/25	400	70	6.5	1.2/2	1/150		DO13
F602	2.4/25	2/25	600	70	6.5	1.2/2	1/150		DO13

CONTROLLED AVALANCHE SILICON DIODES															
Type	V_{RRM}(V)	V_{RA}(V)[I_R=100μA	I_{FRMSM}(A	I_{FSM}(A)/Tj(°C) t=10r	i^2t(A^2S)/T_j(°	I_{FAVM}(A)/T_j(°C	V_{FM}(V)	I_{FM}(A	V_{F0}(V)	r_F(mΩ)	I_R(mA)	T(°C)[V_R=V_R	P_{RSM}(kW	R_{thJC}(°C/W	Case
10A															
D10	400	450...750	15.7	210/25 150/160	110/160	10/125	1.4	35	0.98	10	3	150	30	2.5	DO4
D10	600	720...1000	15.7	210/25 150/160	110/160	10/125	1.4	35	0.98	10	3	150	18	2.5	DO4
D10	800	950...1400	15.7	210/25 150/160	110/160	10/125	1.4	35	0.98	10	3	150	12	2.5	DO4
D10	1000	1200...1700	15.7	210/25 150/160	110/160	10/125	1.4	35	0.98	10	3	150	6	2.5	DO4
16A															
D16	400	450...750	25	250/25 180/175	160/175	16/125	1.4	50	0.82	9.8	3	150	40	2.5	DO4
D16	600	720...1000	25	250/25 180/175	160/175	16/125	1.4	50	0.82	9.8	3	150	25	2.5	DO4
D16	800	950...1400	25	250/25 180/175	160/175	16/125	1.4	50	0.82	9.8	3	150	16	2.5	DO4
D16	1000	1200...1700	25	250/25 180/175	160/175	16/125	1.4	50	0.82	9.8	3	150	10	2.5	DO4
D16	1200	1400...2000	25	250/25 180/175	160/175	16/125	1.4	50	0.82	9.8	3	150	6	2.5	DO4
D16	1400	1600...2400	25	250/25 180/175	160/175	16/125	1.4	50	0.82	9.8	3	150	4	2.5	DO4
25A															
D25	400	450...750	40	400/25 300/160	450/160	25/125	1.4	80	0.9	5.6	5	150	60	1.1	DO5
D25	600	720...1000	40	400/25 300/160	450/160	25/125	1.4	80	0.9	5.6	5	150	35	1.1	DO5
D25	800	950...1400	40	400/25 300/160	450/160	25/125	1.4	80	0.9	5.6	5	150	25	1.1	DO5
D25	1000	1200...1700	40	400/25 300/160	450/160	25/125	1.4	80	0.9	5.6	5	150	13	1.1	DO5
D25	1200	1400...2000	40	400/25 300/160	450/160	25/125	1.4	80	0.9	5.6	5	150	10	1.1	DO5
D25	1400	1600...2400	40	400/25 300/160	450/160	25/125	1.4	80	0.9	5.6	5	150	7	1.1	DO5

ZENER DIODES								
Type	V_Z (V)			I_Z(mA)	r_Z(Ω)	å$_{VzT}$(10^{-4}/°C)	I_{ZM}(mA)	Case
	min.	*nom.*	*max.*					
1W								
PL3V3Z	3.1	3.3	3.5	100	10	-6	285	F126
PL3V6Z	3.4	3.6	3.8	100	10	-5.5	260	F126
PL3V9Z	3.7	3.9	4.1	100	7	-5	240	F126
PL4V3Z	4	4.3	4.6	100	7	-4	215	F126
PL4V7Z	4.4	4.7	5	100	7	-2	200	F126
PL9V1Z	4.8	5.1	5.4	100	5	-1	185	F126
PL5V6Z	5.2	5.6	6	100	2	2.5	165	F126
PL6V2Z	5.8	6.2	6.6	100	2	3.2	150	F126
PL6V8Z	6.4	6.8	7.2	100	2	1	140	F126
PL7V5Z	7	7.5	7.9	100	2	4.5	130	F126
PL8V2Z	7.7	8.2	8.7	100	2	4.8	110	F126
PL9V1Z	8.5	9.1	9.6	50	4	5.1	100	F126
PL10Z	9.4	10	10.6	50	4	5.5	94	F126
PL11Z	10.4	11	11.6	50	7	6	76	F126
PL12Z	11.4	12	12.7	50	7	6.5	79	F126
PL13Z	12.4	13	14.1	50	10	6.5	71	F126
PL15Z	13.8	15	15.6	50	10	7	64	F126
PL16Z	15.3	16	17.1	25	15	7	59	F126
PL18Z	16.8	18	19.1	25	15	7.5	52	F126
PL20Z	18.8	20	21.2	25	15	7.5	47	F126
PL22Z	20.8	22	23.3	25	15	8	43	F126

ZENER DIODES								
Type	$V_Z(V)$			$I_Z(mA)$	$r_Z(\Omega)$	$å_{VzT}(10^{-4}/°C)$	$I_{ZM}(mA)$	Case
	min.	nom.	max.					
1W								
1N3016B	6.4	6.8	7.2	37	3.5	2...6	140	DO13
1N3017B	7	7.5	7.9	34	4	3...7	130	DO13
1N3018B	7.7	8.2	8.7	31	4.5	3.5...7.5	110	DO13
1N3019B	8.5	9.1	9.6	28	5	4...8	100	DO13
1N3020B	9.4	10	10.6	25	7	4...8	94	DO13
1N3021B	10.4	11	11.6	23	8	4...9	86	DO13
1N3022B	11.4	12	12.7	21	9	4...9	79	DO13
1N3023B	12.4	13	14.1	19	10	4...9	71	DO13
1N3024B	13.8	15	15.6	17	14	5...9	64	DO13
1N3025B	15.3	16	17.1	15.5	16	5...9	59	DO13
1N3026B	16.8	18	19.1	14	20	5...9	52	DO13
1N3027B	18.8	20	21.2	12.5	22	5...9	47	DO13
1N3028B	20.8	22	23.3	11.5	23	5...9	43	DO13
1N3029B	22.8	24	25.6	10.5	25	5...9	39	DO13
1N3030B	25.1	27	28.9	9.5	35	5...9	35	DO13
1N3031B	28	30	32	8.5	40	5...9	31	DO13
1N3032B	31	33	35	7.5	45	5...9	29	DO13
1N3033B	34	36	38	7	50	6...10	26	DO13
1N3034B	37	39	41	6.5	60	6...10	24	DO13
1N3035B	40	43	46	6	70	6...10	22	DO13
1N3036B	44	47	50	5.5	80	6...10	20	DO13

SILICON COMMUTATION DIODES										
Type	Absolute Limit Values			Electrical Characteristics						Case
	$V_R(V)$	$I_F(mA)$	$T_j(°C)$	$V_F(V)$	$I_F(mA)$	$I_R(nA)$	$V_R(V)$	$C_{tot}(pF)$	$t_{rr}=t_r(ns)$	
1N4148	75	200	200	1	10	25	20	4	4	DO35
1N4149	75	200	200	1	10	25	20	2	4	DO35
1N4151	50	200	200	1	50	50	20	2	2	DO35
1N4154	25	200	200	1	30	100	25	4	2	DO35
1N4446	75	200	200	1	20	25	20	4	4	DO35
1N4447	75	200	200	1	20	25	20	2	4	DO35
1N4448	75	200	200	1	100	25	20	4	4	DO35
1N4449	75	200	200	1	30	25	20	2	4	DO35
1N4454	50	200	200	1	10	100	20	2	2	DO35
BAY93	20	115	200	1	10	100	10	5	-	DO35
DP450	20	50	125	1	10	0.01	20	1.5	350	DO35
DP451	20	50	125	1	10	0.025	20	1.5	350	DO35

SILICON VARICAP DIODES

Type	Absolute Limit Values		Electrical Characteristics						Case
	V_R(V)	T_j(°C)	C(pF)/V_{R1}(V)		C(pF)/V_{R2}(V)		I_R(μA)/V_R(V)		
BB121	30	150	1.96..2.65	25	–	3	0.	28	DO35
BB122	30	150	2.1...2.8	25	–	3	0.1	28	DO35
BB125	30	150	2...3	25	12	3	1	28	DO35
BB126	30	150	2...3	25	11	3	1	28	DO35
BB139	30	150	4.3...6	25	29	3	0.1	28	DO35
ß310	–	150	–	9	16...28	4	1	12	DO35
ß410	–	150	–	9	7.5...11.5	4	5	12	DO35

IMPATT DIODES

Type	Electrical Characteristics					Bias		Efficiency	Limit Values		Case
	V_{BR}(V) [I_R=1mA	I_R(μA)	C_D(pF)	P_{out}(mW)	f_o(GHz)	I_o(mA)	V_o(V)	n(%) [I_R=I_o	T_j(°C)	R_{thJC}(°C/W)	
BXY0181	70...90	10	0.8	100	8...12	40	95	3	200	30	F–27d
BXY0182	70...90	10	0.8	100	8...12	40	95	3	200	30	F–27d
BXY0301	60...100	10	1	350	10...12	80	100	5	200	23	F–27d
BXY0302	60...100	10	1	350	10...12	80	100	5	200	23	F–27d
BXY0381	60...100	10	1	350	8...10	80	100	5	200	21	F–27d
BXY0372	60...100	10	1	350	8...10	80	100	55	200	21	F–27d
BXY0391	60...100	10	1	350	9...11	80	100	5	200	22	F–27d
BXY0392	60...100	10	1	350	9...11	80	100	5	200	22	F–27d
BXY0501	60...100	10	1.2	500	10...12	100	110	6.5	200	19	F–27d
BXY0502	60...100	10	1.2	500	10...12	100	110	6.5	200	19	F–27d
BXY0581	60...100	10	1.2	500	8...10	100	110	6.5	200	17	F–27d
BXY0582	60...100	10	1.2	500	8...10	100	110	6.5	200	17	F–27d
BXY0591	60...100	10	1.2	500	9...11	100	110	6.5	200	18	F–27d
BXY0522	60...100	10	1.2	500	9...11	100	110	6.5	200	18	F–27d

DOUBLE BASE DIODES (UNIJUNCTION TRANSISTORS)

Type	Case	Absolute Limit Values					Electrical Characteristics							
		P_D(mW)	I_E(mA)	I_E^*(A)	V_{EB1}(V)	V_{B2B1}(V)	n		R_{B2B1}(kΩ)I_E=0		I_{EBO}(μA)	I_{E1}(μA)	I_{E2}(mA)	V_{OB1}(V)
							min.	max.	min.	max.	max.	max.	min.	min.
ROS11	TO5	450	70	2	30	35	0.6	0.8	2.5	10	12	25	1	3
ROS11A	TO5	450	70	2	30	35	0.6	0.8	2.5	10	12	12	1	3
ROS11B	TO5	450	70	2	30	35	0.6	0.8	2.5	10	0.2	6	8	3
ROS11C	TO5	450	70	2	30	35	0.6	0.8	2.5	10	0.02	2	8	3
ROS12	TO5	450	70	2	60	35	0.5	0.75	6.2	12	1	12	8	3
ROS12A	TO5	450	70	2	60	35	0.5	0.6	6.2	12	1	12	8	3
ROS12B	TO5	450	70	2	60	35	0.56	0.68	6.2	12	1	12	8	3
ROS12C	TO5	450	70	2	60	35	0.62	0.75	6.2	12	1	12	8	3
2N1671	TO5	450	50	2	30	35	0.47	0.62	4.7	9.1	12	25	8	–
2N1671A	TO5	450	50	2	30	35	0.47	0.62	4.7	9.1	12	25	8	3
2N1671B	TO5	450	50	2	30	35	0.47	0.62	4.7	9.1	0.2	6	8	3
2N1671C	TO5	450	50	2	30	35	0.47	0.62	4.7	9.1	0.02	2	8	3
2N2160	TO5	450	50	2	30	35	0.47	0.8	4	12	12	25	8	3
2N2646	TO18	300	50	2	30	30	0.56	0.75	4.7	9.1	12	5	4	3
2N2647	TO18	300	50	2	30	30	0.68	0.82	4.7	9.1	0.2	2	8	6
2N3479	TO5	450	50	2	30	35	0.47	0.62	4.7	9.1	12	20	6	6
2N3480	TO5	450	50	2	30	35	0.56	0.75	4.7	9.1	12	20	4	6
2N3481	TO5	450	50	2	30	35	0.7	0.85	4.7	9.1	12	20	4	6

LOW & MEDIUM POWER THYRISTORS												
Type	VFRM /VRRM (V	IFRMSM(A)	IFSM(A)/Tj(°C	$i^2t(A^2s)/Tj=100$°	di/dt(A/µs)	dv/dt(V/µs)	tq/T_j(µs/°C	$I_F=I_R$(mA)/T_j(°C	V_{GT}(V)	I_{GT}(mA)	R_{thJC}(°C/W	Case
			t=10ms	t=10ms				$V_F=V_{FRM}$ $V_R=V_{RRM}$	Tc=25°C	Tc=25°C		
1A												
T1N05	50	1.57	10 100	0.5	20	50	40 100	1 120	2	10	30	SOT32
T1N1	100	1.57	10 100	0.5	20	50	40 100	1 120	2	10	30	SOT32
T1N2	200	1.57	10 100	0.5	20	50	40 100	1 120	2	10	30	SOT32
T1N4	400	1.57	10 100	0.5	20	50	40 100	1 120	2	10	30	SOT32
T1N6	600	1.57	10 100	0.5	20	50	40 100	1 120	2	10	30	SOT32
T1N8	800	1.57	10 100	0.5	20	50	40 100	1 120	2	10	30	SOT32
T1R105	50	1.57	10 100	0.5	20	50	10 100	1 120	2	10	30	SOT32
T1R1	100	1.57	10 100	0.5	20	50	10 100	1 120	2	10	30	SOT32
T1R2	200	1.57	10 100	0.5	20	50	10 100	1 120	2	10	30	SOT32
T1R4	400	1.57	10 100	0.5	20	50	10 100	1 120	2	10	30	SOT32
T1R6	600	1.57	10 100	0.5	20	50	10 100	1 120	2	10	30	SOT32
T1R8	800	1.57	10 100	0.5	20	50	10 100	1 120	2	10	30	SOT32
3A												
T3N05	50	4.7	30 100	4.5	50	50	30 25	1.5 125	3	30	4	TO66(F22)
T3N1	100	4.7	30 100	4.5	50	50	80 25	1.5 125	3	30	4	TO66(F22)
T3N2	200	4.7	30 100	4.5	50	50	80 25	1.5 125	3	30	4	TO66(F22)
T3N4	400	4.7	30 100	4.5	50	50	80 25	1.5 125	3	30	4	TO66(F22)
T3N6	600	4.7	30 100	4.5	50	50	80 25	1.5 125	3	30	4	TO66(F22)

DIACS								
Type	V_{BO1} / V_{BO2} (V		I_{BO1} / I_{BO2} (µA)	V_{BOmax}(V)	V_{L1} / V_{L2} (V)	I_{TRM} (A)	P_D(W)	Case
	min.	max.	max.	max.	min.	t_p=20µs f=100Hz		
DC32	28	36	300	4	5	1.5	0.3	F126
DC32A	28	36	50	2	5	1.5	0.3	F126
DC38	34	42	300	4	5	1.5	0.3	F126
DC38A	34	42	50	2	5	1.5	0.3	F126
DC44	40	48	300	4	7	1.5	0.3	F126
DC44A	40	48	50	2	7	1.5	0.3	F126
DC50	46	54	300	4	7	1.5	0.3	F126
DC50A	46	54	50	2	7	1.5	0.3	F126
index T corresponds to index F								

TRIACS														
Type	V_{TRM}(V)	I_{TRMSM}(A)	I_{TSM}(A)	$i^2t(A^2s)$	V_{TM}(V)	I_{TM}(A)	V_{TO}(V)	r_T(mΩ)	dV/dt(V/µs)	i_T(mA)	V_G(V)	I_G(mA)	R_{thJC}(°C/W)	Case
		T_c=80°C	t=10ms	t=10ms						$V_T=V_{TRM}$				
TB6N2	200	6	48	11.5	2.3	8.5	1.8	52	5	0.5	2.5	100	3	TO220
TB6N3	300	6	48	11.5	2.3	8.5	1.8	52	5	0.5	2.5	100	3	TO220
TB6N4	400	6	48	11.5	2.3	8.5	1.8	52	5	0.5	2.5	100	3	TO220
TB6N5	500	6	48	11.5	2.3	8.5	1.8	52	5	0.5	2.5	100	3	TO220
TB6N6	600	6	48	11.5	2.3	8.5	1.8	52	5	0.5	2.5	100	3	TO220
TB10N2	200	10	80	32	1.9	1.4	1.18	38	5	0.5	2.5	100	2.4	TO220
TB10N3	300	10	80	32	1.9	1.4	1.18	38	5	0.5	2.5	100	2.4	TO220
TB10N4	400	10	80	32	1.9	1.4	1.18	38	5	0.5	2.5	100	2.4	TO220
TB10N5	500	10	80	32	1.9	1.4	1.18	38	5	0.5	2.5	100	2.4	TO220
TB10N6	600	10	80	32	1.9	1.4	1.18	38	5	0.5	2.5	100	2.4	TO220
index T corresponds with index F														

SILICON PHOTODIODES					
Type	***S****(μA/klx)*	***V_R****(V)*	***I_D****(nA)*	***t_r****(μs)*	*Case*
ROL21	30	75	20	1	7
ROL021	30	50	20	1	6
ROL22	7	70	20	1	8

SOLAR CELLS								
Type	*Electrical Characteristics* T_A*=25°C AMO*						*Absolute Limit Values*		
	$V_O(V)$		$I_{SC}(A)$		$P_M(W)$		$E(kW/m^2)$	$T_j(°C)$	$T_{storage}(°C)$
	min	*typ*	*min*	*typ*	*min*	*typ*			
Cs-01-C2	0.5	0.56	0.5	0.55	0.25	0.3	1	150	-40...100
Cs-01-C3	0.5	0.56	1.2	1.3	0.5	0.52	1	150	-40...100
Cs-01-P2	0.5	0.56	0.7	0.75	0.3	0.32	1	150	-40...100
Cs-01-S2	0.5	0.56	1.05	1.15	0.4	0.45	1	150	-40...100

AMO = 100mW/cm^2 in solar spectrum

LIGHT EMITTING DIODES (LEDs) *GaAsP*					
Type	I_v*(mcd)*	I_F*(mA)*	*Color*	t_r*(ns)*	*Case*
ROL02	0.6	20	red	15	6
ROL05	0.8	20	orrange	15	6
ROL07	0.8	20	yellow	15	6

STANDARD LIGHT EMITTING DIODES								
Type	*Optoelectrical Characteristics* T_A*=25°C* I_F*=20mA*							
	I_V*(mcd)*		*Wave Length of max eff(nm)*		*Vis Angle(grad)*	V_F*(V)*	V_{BR}*(V)*	C_o*(pF)*
	min	*max*	*min*	*max*			I_R*=100μA*	V_F*=0V f=1MHz*
MDE1101R	0.1	1	645	680	40	2	3	70
MDE1102R	1	2	645	680	40	2	3	70
MDE1103R		2	645	680	40	2	3	70
MDE1101P	0.3	1.5	625	640	40	3	3	60
MDE1102P	1.5	2.5	625	640	40	3	3	60
MDE1103P		2.5	625	640	40	3	3	60
MDE1101G	0.3	1.5	573	590	40	3	3	60
MDE1102G	1.5	2.5	573	590	40	3	3	60
MDE1103G		2.5	573	590	40	3	3	60
MDE1101V	0.1	1	554	570	40	3	3	60
MDE1102V	1	2	554	570	40	3	3	60
MDE1103V		2	554	570	40	3	3	60
	Absolute Limit Values							
	I_F*(mA)*		I_{peak} *(A)*		P_D*(mW)*	T_A*(°C)*	$T_{storage}$*(°C)*	$T_{soldering}$*(°C)*
		50		1	150	0...+70	-40...+85	260

Note: R=red V=green G=yellow P=amber

LOW POWER LOW FREQUENCY SILICON pnp TRANSISTORS													
Type	Absolute Limit Values					Electrical Characteristics							
	V_{CEO}(V)	V_{EBO}(V)	I_C(mA)	P_D(mW)	T_j(°C)	V_{CEsat}(V)	I_{Csat}(mA)	h_{21e}(h_{21E}	I_C(mA)	V_{CE}(V)	f_T(MHz)	F(dB)	f(kHz)
BC177	45	5	100	300	175	0.95	100	75-260	2	5	200	10	1
BC178	25	5	100	300	175	0.95	100	75-260	2	5	200	10	1
BC179	20	5	100	300	175	0.95	100	125-500	2	5	200	4	0.03-15
BC250	20	5	100	300	150	0.3	10	35-600	1	1	180	10	1
BC251	45	5	100	300	150	0.3	10	125-900	2	5	200	10	1
BC252	25	5	100	300	150	0.3	10	125-900	2	5	200	10	1
BC253	25	5	100	300	150	0.3	10	125-900	2	5	200	4	0.03-15
BC256	64	5	100	300	150	0.3	10	125-500	2	5	200	10	1
BC327	45	5	800	625	150	0.7	500	100-630	100	1	100	-	-
BC328	25	5	800	625	150	0.7	500	100-630	100	1	100	-	-

LOW POWER LOW FREQUECNY SILICON npn TRANSISTORS													
Type	Absolute Limit Values					Electrical Characteristics							
	V_{CEO}(V)	V_{EBO}(V)	I_C(mA)	P_D(mW)	T_j(°C)	V_{CEsat}(V)	I_{Csat}(mA)	h_{21e}(h_{21E}	I_C(mA)	V_{CE}(V)	f_T(MHz)	F(dB)	f(kHz)
BC107	45	6	100	300	175	0.6	100	125-500	2	5	300	10	1
BC108	20	5	100	300	175	0.6	100	125-900	2	5	300	10	1
BC109	20	5	100	300	175	0.6	100	240-900	2	5	300	4	0.05-15
BC170	20	5	100	300	150	0.4	30	35-800	1	1	100	10	1
BC171	45	6	100	300	150	0.6	100	125-900	2	5	300	10	1
BC172	25	5	100	300	150	0.6	100	125-900	2	5	300	10	1
BC173	25	5	100	300	150	0.6	100	240-900	2	5	300	4	0.03-15
BC174	64	5	100	300	150	0.6	100	125-500	2	5	300	10	1
BC190	64	5	100	300	175	0.6	100	125-500	2	5	300	10	1
BC237	45	6	100	300	150	0.6	100	125-500	2	5	300	10	1
BC238	25	5	100	300	150	0.6	100	125-900	2	5	300	10	1
BC239	25	5	100	300	150	0.6	100	240-900	2	5	300	4	0.03-15
BC337	45	5	800	625	150	0.7	500	100-630	100	1	100	-	-
BC338	25	5	800	625	150	0.7	500	100-630	100	1	100	-	-

MEDIUM POWER LOW FREQUENCY SILICON TRANSISTORS													
Type	Absolute Limit Values					Electrical Characteristics				T_{amb}=25°C			
	V_{CEO}(V)	V_{EBO}(V)	I_C(A)	P_D(W)	T_j(°C)	V_{CEsat}(V)	I_{Csat}(mA)	f_T(MHz)	I_C(mA)	I_{CBO}(mA)	V_{CE}(V)	h_{21E}	I_C(mA)
npn													
BD135	45	5	1.5	6.5	150	0.6	500	50	50	100	30	40-250	150
BD137	60	5	1.5	6.5	150	0.6	500	50	50	100	30	40-160	150
BD139	80	5	1.5	6.5	150	0.6	500	50	50	100	30	40-160	150
pnp													
BD136	45	5	1.5	6.5	150	0.6	500	50	50	100	30	40-250	150
BD138	60	5	1.5	6.5	150	0.6	500	50	50	100	30	40-160	150
BD140	80	5	1.5	6.5	150	0.6	500	50	50	100	30	40-160	150

POWER LOW FREQUENCY SILICON TRANSISTORS												
Type	Absolute Limit Values							Electrical Characteristics				
	V_{CBO}(V)	V_{CEO}(V)	V_{EBO}(V)	I_C(A)	P_D(W)	T_j(°C)	R_{thjc}(°C/W)	V_{CEsat}(V)	I_{Csat}(A)	h_{21E}	I_C(A)	f_T(MHz)
npn												
2N3055	100	60	7	15	117	200	1.5	1.1	4	20-70	4	1.5

FIELD EFFECT TRANSISTORS										
Type		*Absolute Limit Values*				*Electrical Characteristics*				
		$V_{DS}(V)$	$V_{GS}(V)$	$I_D(mA)$	$P_D(mW)$	$V_{thr}(V)$	$I_{D0}(mA)$	$I_{GS}(nA)$	$r_{DS}[ON](\Omega)$	$g_m(mA/V)$
MOS-FET	channel									
ROS01	**p**	−30	−40	30	200	−4	1	0.05	100	9
ROS02	**n**	30	40	30	200	2...10	1...3	0.01	500	5
ROS04	**p**	−25	−	30	200	−2...−5	0.01	−	350	1
ROS05	**p**	−30	−40	30	200	−4	1	0.05	100	5
2X ROS01	double									
J-FET	channel									
ROS104	**p**	−25	25	30	200	3.5	1	0.1	250	4
BFW10	**n**	−	−30	−	−	−8	8...20	−0.1	−	−
BFW11	**n**	−	−30	−	−	−6	4...10	−0.1	−	−
2N4091	**n**	40	−40	−	1800	−5...10	30	−	30	−
2N4092	**n**	40	−40	−	1800	−2...7	15	−	30	−
2N4093	**n**	40	−40	−	1800	−1...5	8	−	30	−

BIBLIOGRAPHY / REFERENCES

*DIODES AND TRANSISTORS
BIBLIOGRAPHY/REFERENCES

"MINIWATT" TRANSISTORS AND DIODES 1957

BEBOP TO THE BOOLEAN BOOGIE - CLIVE MAXFIELD 1957

CIRCUITE CU TRANZISTOARE IN INDUSTRIE - I.FELEA S.A. 1964

BAZELE PROIECTARII CALCULATOARELOR NUMERICE - YAOHAN CHU 1968

CIRCUITE CU SEMICONDUCTOARE IN INDUSTRIE - A. VATASESCU S.A.1971

MANUALUL RADIOAMAORULUI - M. TANCIU, I. VIDRASCU 1971

TRANZISTOARE CU EFECT DE CAMP - I. RISTEA, M. BODEA 1972

BAZELE ELECTRONICII MODERNE - P.E.GRAY, C.I. SEARLE 1973

TECHNICA TELEVIZIUNII IN CULORI - G. RAYMOND 1971

MULLARD TECHNICAL BOOK - BOOK ONE PART THREE 1975

DE LA EFECTUL FOTOELECTRIC LA CELULA SOLARA - M. NEGREANU, GH. BALUTA 1981

AGENDA RADIOELECTRONISTULUI - N. DRAGULANESCU 1984

ELECTROTECHNICA DE LA A LA Z - E. MICU 1985

DIODE CU SILICIU – CATALOG - P.A. DAN, D.M. LUCA, A. ALBU, T. DUNCA, G. PRIMEJDIE 1986

DISPOZITIVE SEMICONDUCTOARE – CATALOG CONDENSAT IPRS

SILICON CONTROLLED RECTIFIER MANUAL – GENERAL ELECTRIC COMPANY
EDITOR F.W. GUTZWILLER

THE ART OF ELECTRONICS - P. HOROWITZ, W. HILL 1989-2001

HANDBOOKOF ELECTRICAL AND ELECTRONICS TECHNOLOGY- CURTIS D. JOHNSON 1996

AN INTRODUCTION TO DIODES AND TRANSISTORS - EDWARD M. CHERRY 1998

COMPLETE GUIDE TO SEMICONDUCTOR DEVICES - KWOK K. NG 2002

LES DIODES SEMICONDUCTRICES - B. BOITTIAUX 2004

LE TRANSISTOR BIPOLAIRE - B. BOITTIAUX 2004

RF SYSTEMS, COMPONENTS AND CIRCUITS HANDBOOK 2005

AN INTRODUCTIOB TO SEMICONDUCTOR DEVICES - DONALD NEAMEN 2006

PRACTICAL ELECTRONICS - ANDY COOPER 2016

TEACH YOURSELF ELECTRICITY AND ELECTRONICS - STAN GIBILISCO, SIMON MONK 2016

** Revised edition. First appearance: DIODE SI TRANZISTOARE, Editura Technica, Bucuresti 1987*

www.ingramcontent.com/pod-product-compliance
Lightning Source LLC
LaVergne TN
LVHW071201210926
848482LV00014B/704